QK
711
.H2
1969
Hales
Vegetable staticks
AUG JAN 1992 '71
MAR '87
OCT '75
OCT '82
39359
KVCC
KALAMAZOO VALLEY
COMMUNITY COLLEGE
LIBRARY
WITHDRAWN

History of Science Library : Primary Sources

Editor : Michael A. Hoskin
Lecturer in the History of Science, Cambridge University

VEGETABLE STATICKS

History of Science Library

Editor: Michael A. Hoskin
Lecturer in History of Science, Cambridge University

The Conflict between Atomism and Conservation Theory, 1644–1860
W. L. Scott

The Earth in Decay
A History of British Geomorphology, 1578–1878
G. L. Davies

Theories of Light
from Descartes to Newton
A. I. Sabra

Medicine in Medieval England
C. H. Talbot

The Origins of Chemistry
R. P. Multhauf

The English Paracelsians
A. G. Debus

A History of the Theories of Rain
W. E. Knowles Middleton

William Herschel and the Construction of the Heavens
M. A. Hoskin

History of Science Library: Primary Sources

In preparation

Fundamenta Medicinae
Friedrich Hoffman
Translated and introduced by L. S. King

A Philosophicall Key
Robert Fludd
Edited and introduced by A. G. Debus

Vegetable Staticks

STEPHEN HALES

Foreword to this edition by

M. A. HOSKIN

Lecturer in the History of Science,
Cambridge University

MACDONALD: LONDON

AND

AMERICAN ELSEVIER INC.: NEW YORK

First published 1727

This edition first published by
The Scientific Book Guild, August 1961

Reissued in the History of Science Library by
Macdonald & Co. (Publishers) Ltd, 1969

British Edition published by
Macdonald & Co. (Publishers) Ltd
49/50 Poland Street
London W.1

American Edition published by
American Elsevier Publishing Company, Inc.
52 Vanderbilt Avenue
New York N.Y. 10017

Standard Book Numbers
British SBN 356 03012 1 (cloth-bound edition)
British SBN 356 03013 X (paper-bound edition)
American SBN 444 19704 4 (cloth-bound edition)

Library of Congress Catalog Card Number 75-101421

Printed in Great Britain by
Hazell Watson & Viney Ltd,
Aylesbury, Bucks

PREFACE

The History of Science Library: Primary Sources

In its early years, the History of Science Library consisted solely of new books written specially for the Library. But the first announcements of the Library promised also editions of "primary sources"—the documents on which the historian bases his interpretation of history—although it was never the intention to put out yet another collection of facsimile reprints. Instead, a series of primary sources has been planned within the general framework of the Library, each of which will be published for a purpose and will be presented in the way that best serves that purpose. Some texts will be translated for the first time, others edited from unpublished manuscripts, and others reset in modern type for greater legibility. We hope in this way to put into the hands of students and scholars material which might not otherwise be easily available to them.

M. A. Hoskin

FOREWORD

The Scientific Revolution that reached its climax in the seventeenth century is perhaps the most important yet at the same time one of the most intriguing episodes in the whole history of science. It is important because from it emerged science with the aims and methods that we know today; intriguing because its many tangled threads have yet to be unravelled in a way that a majority of historians will accept. From one standpoint, the Scientific Revolution meant the rejection of a primarily biological view of nature and its replacement by a mathematical and mechanical view. At least as far back as the Pythagoreans, the world had been looked on as a cosmos, a living and integrated whole, and this outlook was reinforced by the Aristotelian insistence on purpose in nature. The most obvious embodiment of these ideas was to be found in the analogy between the microcosm, the individual living body, and the macrocosm, the world as a whole.

From the late Middle Ages, however, different trends began to emerge. Clocks and other machinery of great complexity were constructed; ancient writings setting out a materialist philosophy of nature were read again; and the development of printing made possible once more the advance of technical mathematics. Gradually the clarity and intelligibility of the explanation which a clockmaker might give of a clock began to seem an ideal for explanations of the natural world. As Kepler put it, "My goal is to show that the heavenly machine is not a kind of divine living being but is similar to a clockwork".

As the seventeenth century wore on this movement gathered strength, and by the time the Royal Society was founded in 1660 the mechanical philosophy, the view that at least physical and chemical changes were to be explained solely in terms of matter in motion, seemed so unexceptional, so natural, that it was no longer regarded merely as one of several possible approaches to the natural world; no other view could be taken seriously, and so the early Fellows had no hesitation in adopting it in spite of their resolve not to accept any scientific system until they had conclusive experimental evidence. Francis Bacon, their inspiration in so many ways, had done the same thing for the same reason earlier in the century, and on the continent Descartes had gone further and made matter-in-motion the basis of his explanations of changes not only in inanimate matter but in living things as well. There was, he held, no essential difference between animals and the automata which had been a source of fascination since ancient times; there was a difference in the complexity of their organisations, but no sharp distinction was to be drawn between living and non-living, or between the natural and the artefact.

Whether or not one was prepared to go all the way with Descartes, there was everything to be gained by studying animals and plants in physical (and perhaps chemical) terms. As Stephen Hales was to put it in 1727 in the Introduction to *Vegetable Staticks,*

> Since we are assured that the all wise Creator has observed the most exact proportions, *of number, weight and measure,* in the make of all things; the most likely way therefore, to get any insight into the nature of those parts of the creation, which come within our observation, must in all reason be to number, weigh and measure.

This was the essence of (in Hales's term) "the statical way" of investigating nature. But it was one thing to say "we may look upon a tree as a complicated engine", or to admire the "variety of masterly strokes of machinery" in plants; it was quite another to put this programme into practice, to simplify and idealise complex biological problems so that the statical way could be applied in science.

In zoology the breakthrough was achieved by William Harvey in his work *On the Motion of the Heart and Blood in Animals* (1628). By viewing the heart and blood vessels as an hydrodynamical system of pumps, pipes and valves, Harvey was able to isolate one fundamental element in the complex of problems surrounding nutrition and respiration along with the many associated organs and vessels. By making a minimal estimate of the *quantity* of blood crossing from one side of the heart to the other, Harvey was able to show that this quantity, even though below the true figure, was so great that it could not possibly be either produced or used by the body. In other words, it must be the same blood that was crossing the heart time and again.

Harvey's work, like so many of the classics of science, answered one question only to raise others. Christopher Wren investigated the effect of replacing blood by other fluids; Richard Lower extended Harvey's account and showed that the blood changed colour in the lungs; Robert Hooke devised a technique for ventilating the lungs artificially; experiments in blood transfusion were carried out in London and on the continent; and Hales himself, in experiments which he mentions in *Vegetable Staticks* but describes in full in his *Haemastaticks* (1733), studied the pressure of the blood and related problems. Meanwhile, on the Continent, Giovanni Borelli had discussed in *De Motu Animalium* (1680–81) the mechanics of animal movement in terms of levers and forces. These and other similar works made permanent advances in the understanding of animal physiology, even if gradually it became clear that on many problems in this field a more subtle attack was required.

In botany, however, the new techniques were far less easy to apply, and it was almost exactly a century after the appearance of Harvey's classic that a comparable breakthrough occurred with the publication of *Vegetable Staticks*. True, just as Harvey had not been the first to apply a mechanical approach in zoology (earlier in the seventeenth century a colleague of Galileo at Padua, Santorio Santorio, had weighed himself in a variety of circumstances, to say nothing of Galen's partially quantitative study of the heart structure in Roman times), so Hales was not the first to "number, weigh and measure" in botany. Johann Baptista van Helmont, a contemporary of Harvey, acting on a suggestion made in the fifteenth century by Nicholas of Cusa, had grown a tree in a pot under carefully controlled conditions, supplying it only with water, and (knowing nothing of the rôle of the air in plant growth) had concluded that the increase in the weight of the plant was wholly due to the incorporation of the water it had received. But Hales was the first to carry through a planned series of experiments resulting in permanent and important additions to knowledge, posing new problems and illustrating techniques by which they might be solved. His account, a classic of plant physiology, is now reprinted for the first time since the eighteenth century.

Stephen Hales was born in 1677 into a prominent Kentish family. He entered Bene't College (Corpus Christi College), Cambridge, in 1696, the year in which Newton left the university to become Warden of the Mint in London. Neither Newton's long tenure of the Lucasian professorship nor the publication of his *Principia* in 1687 had had much impact on the university. In fact it was the rival system of Descartes, the one version of the mechanical philosophy which could be taught systematically thanks to the textbook of Jacques Rohault, which was widely accepted. This, even the most partisan Newtonians agreed, was a big step forward from the Aristotelianism of earlier years. A defect in the teaching of science more serious than the indifference to Newtonianism was the absence of experiments. The pioneers of the Royal Society were deeply convinced that it was by the "experimental philosophy" that knowledge of nature would grow. As Hales was to put it,

> The wonderful and secret operations of Nature are so involved and intricate, so far out of the reach of our senses, as they present themselves to us in their natural order, that it is impossible for the most sagacious and penetrating genius to pry into them, unless he will be at the pains of analysing Nature, by a numerous and regular series of Experiments.

But the universities were not centres of research, and experiments were not required for the instruction of undergraduates. Individuals might research and experiment in private, as Newton had done in showing that white light was not simple but composed of colours; but such individuals were rare, as Newton had discovered when he tried to found a research club in Cambridge. "That which chiefly dasht the buisiness", he wrote in 1685, "was the want of persons willing to try experiments".

Hales spent thirteen years in Cambridge before leaving to take charge of the quiet riverside parish of Teddington shortly after he was ordained in June, 1709. It was an eventful period for Cambridge science. Cartesianism was still firmly entrenched when Hales left, but now Newtonian ideas were being widely canvassed and there were frequent criticisms of Descartes even in the very notes to the standard edition of Rohault's textbook. This was due in part to the appearance in 1704 of Newton's long-awaited *Opticks*. Unlike the forbiddingly mathematical *Principia*, which was a closed book to all but a handful of specialists, the *Opticks*, intended to prove the properties of light "by reason and experiments", could be read by experimenters whether mathematicians or not. More exciting still, Newton, who was gifted with an insight into the workings of nature going far beyond what he could establish by proof, had given some hints in the *Opticks* of what was at the back of his mind. He had done this quite openly, in the form of "Queries", and these grew in number as edition succeeded edition. As a consequence, the *Opticks* exercised an immense influence in the eighteenth century, and its impact on Hales is typical. In *Vegetable Staticks* we meet quotations from the Queries time and again—in fact Hales adopts the same literary device himself—and when he turns from the question "in what manner does such and such happen?" to the question "by what means does this come about?" we find that he develops the Newtonian programme whereby all (or almost all) changes are to be explained by the action of gravitational attraction and other attractive and repulsive forces.

All this Hales learned at Cambridge, but it was part of the more academic approach to science. He was also among the first to benefit from the half-way house between instruction and research that was provided by the introduction of the lecture-demonstration. In 1706 the redoubtable Richard Bentley, Master of Trinity College, secured the election of Roger Cotes as first Plumian professor of astronomy, and provided him with an observatory over the main gate of Trinity. He then installed William Whiston, Newton's successor, near by, and Whiston and Cotes soon joined forces to give a course on hydrostatics and pneumatics which they illustrated with experiments. Lecture-demonstrations such as these introduced the more lively minded undergraduates to the use of experiments and did so with a system and purpose that was often lacking in the Royal Society meetings. Another fine opportunity of which Hales took advantage occurred when in 1703 John Francis Vigani of Verona, after years of teaching chemistry privately in Cambridge by means of experiments, was established in a professorship. Bentley promptly provided him with a laboratory in Trinity College, and Hales mentions taking part in an experiment there on p. 112 of *Vegetable Staticks*.

"Lively minded" is altogether insufficient to describe William Stukeley, who descended upon Hales's college in 1703. After two years he was given a special room in college

> to dissect in, & practise Chymical Experiments, which had a very strange appearance with my Furniture in it, the wall was generally hung round with Guts, stomachs, bladders, preparations of parts & drawings. I had sand furnaces, Calots, Glasses, & all sorts of Chymical Implements. . . . Here I & my Associats often dind upon the same table as our dogs lay upon. I often prepared the pulvis fulminans & sometime surprizd the whole College with a sudden explosion.

Stukeley's sense of humour was sometimes macabre:

> Ashwensday 1708. We took up old Hoyes that hangd himself & was buryed in the highway, & dissected him, & afterwards made a sceleton of his bones, & put them in a fine Glass case with an inscription in Latin.

Hales, a much more sober character and ten years Stukeley's senior, was caught up in this whirlwind of activity. Together they learned physics and chemistry from the university professors and did some experimenting on their own account; together, on Stukeley's initiative, they explored the countryside and collected and dissected, and so Hales learned the rudiments of biology. Stukeley was studying medicine, and his friend soon became interested in experiments on living animals. Writing in 1727, Hales tells us that

> About 20 years since, I made several haemastatical Experiments on Dogs, and 6 Years afterwards repeated the same on Horses and other Animals, in order to find out the real force of blood in the Arteries.

In 1707 he was still in Cambridge with Stukeley, officially studying Divinity; "six years afterwards" he was Perpetual Curate of Teddington. Yet he did not regard these experiments, or any part of his scientific activity, as diverting him from his principal studies or, later, from his care of souls. As he grew older he gave more and more attention to work which would bring immediate practical benefits—in *Vegetable Staticks* we already find this emphasis on many occasions—but always, for him as for so many of his scientific contemporaries, to study nature was to be

> sure of entertaining our minds after the most agreeable manner, by seeing in every thing, with surprising delight, such plain signatures of the wonderful hand of the divine Architect, as must necessarily dispose and carry our thoughts to an act of adoration, the best and noblest employment and entertainment of the mind.

Science was a religious pursuit, and in *Vegetable Staticks* our attention is constantly drawn to what Hales sees as the design and purpose in nature.

Even so, the paradox of the kindly Vicar performing these horrifying experiments on blood pressure in animals of which we have a taste on p. 61 of *Vegetable Staticks* appalled and puzzled his friends. His Catholic neighbour Alexander Pope is reported as saying "he commits most of these barbarities with the thought of being of use to man. But how do we know that we have a right to kill creatures that we are so little above as dogs, for our curiosity?"; and Thomas Twining wrote:

> Green Teddington's serene retreat
> For Philosophic Studies meet,
> Where the good Pastor Stephen Hales
> Weighed moisture in a pair of scales,
> To lingering death put Mares and Dogs,
> And stripped the Skins from living Frogs.
> Nature he loved, her Works intent,
> To search or sometimes to torment.

Hales remained at Teddington for over fifty years, right up to his death in 1761 at the age of eighty-three. As Vicar he proved upright and kind if somewhat strict to the sinners of the parish, whom he required to do public penance. When he first arrived he was little known in public circles, but by 1717 his work with animals had brought him to the notice of the Royal Society, then and for many years to come an essentially amateur body, and the following year he was elected Fellow. It was about then that he began his researches in plant physiology, and in 1727, when *Vegetable Staticks* appeared, Hales was elected to the Council of the Royal Society. His increasing prominence in the Society had brought him into more regular contact with physicians who, because of the excellence of the Royal College of Physicians, had always been an important minority in the Society. He was now becoming aware of the medical implications of his work and, having acquired a considerable experimental knowledge of chemistry, he tried to see if he could discover some chemical means of relieving the "distemper of the stone", renal and vesical calculus. He was unsuccessful, although he did devise a useful instrument for the same purpose. His results he described in full, along with his experiments on blood-pressure, in *Haemastaticks* (1733). The second edition of *Vegetable Staticks* had appeared in 1731, and the two books were combined under the general title of *Statical Essays*. A third edition of *Statical Essays* appeared in 1738 and the following year Hales was awarded the Copley medal of the Royal Society, apparently for his relatively worthless work on calculus. In 1753 Hales was elected one of the eight foreign members of the French Academy of Sciences, and by the time the fourth edition of *Statical Essays* appeared in 1769, *Vegetable Staticks* had been translated into Dutch, French, German and Italian. But with his increasing public prominence (in 1733 he became an Oxford D.D., and towards the end of his life he became a royal chaplain and was offered a Canonry of Windsor) his interests were turning more and more away from pure science and towards matters of immediate benefit to mankind. He pamphleteered, with prudent anonymity, against the unrestricted sale of "distilled spirituous liquors, the bane of the nation", the effects of which we know so well from William Hogarth. He was appointed by Parliament one of the "trustees" to investigate the claim of Joanna Stephens to have invented a cure for the stone, and with less than his usual scientific caution agreed that she should be paid the £5000 reward. He became a Trustee of the Colony of Georgia, which gave him worries and troubles for over twenty years. He interested himself in the problems of providing healthy conditions on ships, and especially in the difficulties in the way of adequate ventilation; the Admiralty adopted his windmill device for fanning air, though only after many years of fluctuating fortunes in rivalry with another invention by one Samuel Sutton; in other quarters Hales had more influence and his device was used in emigrant and slave ships as well as in prisons, hospitals and mines. A widower after only a year of marriage, Hales became a venerable and respected public figure, alert in mind and body up to the short illness which led to his death in 1761.

Vegetable Staticks, which with *Haemastaticks* forms the basis of Hales's scientific reputation, contains the fruit of his study of plant physiology in the years between 1718, when at the Royal Society he "informed ye President that he had lately made a new Experiment upon the Effect of ye Sun's warmth in raising ye Sap in trees" and 1725, when at another meeting "The Revd. Mr. Hales communicated a Treatise concerning the Power of Vegetation consisting of six heads of experiments delivered in six chapters". But if in 1725 his botanical work was well advanced some important research still remained to be done before his investigations could be published. As he writes in the Preface,

> Finding by many Experiments in the fifth chapter, that the Air is plentifully inspired by Vegetables, not only at their roots, but also thro' several parts of their Trunks and Branches; this put me upon making a more particular inquiry into the nature of the Air; and to discover, if possible, wherein its great importance to the life and support of Vegetables might consist; on which account I was obliged to delay the publication of the rest of these Experiments, which were read two years since before the Royal Society, till I had made some progress in this inquiry. An account of which I have given in the sixth chapter.

Early in 1727 his chemical experiments were completed and read to the Royal Society at successive meetings. Sir Isaac Newton, now near the end of his long life, ordered the papers to be printed, and it is his *imprimatur* which appears inside the title page.

Vegetable Staticks displays all the essential elements of Hales's scientific thought: his application of the "statical way of inquiry", his deep reverence for the work of "the divine Architect", his alertness to possible practical applications of his work, his interpretation of the underlying mechanism in Newtonian terms, above all his mature use of controlled experimental techniques. Hales makes it all look so easy; only occasionally, as on p. 80, where he admits his failure to carry through an experiment reported by another writer, or on pp. 87 and 119, where he describes with unusual care the apparatus he is using, do we have a hint of the many disappointments he must have endured.

Hales begins by examining the quantities of moisture taken in by the roots and "perspired" or, as we would say, "transpired", by the leaves. The motion of the sap had been studied as long ago as 1669 by John Ray, whose *Catalogue* of Cambridgeshire plants had been carried by Stukeley and Hales on their rambles; but Hales brings the power of numbers to his proofs that a plant takes in large quantities of water through its roots provided that it has leaves from which the water may evaporate; and that without the leaves the plant will take in only a little moisture, even when the roots are embedded in the earth and even when the water is introduced into the stem under pressure. In this way Hales was led to study the importance of the leaves and to consider other possible rôles they might fill, rôles which we now know to be of fundamental importance. In the last chapter he remarks that "the leaves seem also designed for many other noble and important services", and compares the leaves to lungs in animals, "plants very probably drawing thro' their leaves some part of their nourishment from the air". And in a famous passage he asks in the manner of a Newtonian Query: "May not light also, by freely entering the expanded surfaces of leaves and flowers, contribute much to the ennobling the principles of vegetables?" This has been interpreted as a glimpse of the process of photosynthesis, by which the plant uses the energy from the sun to incorporate carbon from the air, although for Hales light is a substance rather than a source of energy. But this is to anticipate. Hales meanwhile goes on to examine specimens of sap, and to study the moisture and temperature of the earth; and we may note here the difficulties caused by the lack of an agreed thermometric scale. He then gives a typical explanation in mechanical terms for the phenomena he has described.

In Chapter II Hales discusses the actual pressure with which the sap is moved. As he has told us in the Preface, after being in despair how to measure this pressure he hit upon a possible method when he accidentally cut a vine during the bleeding season; he tried to stop the sap flowing by binding a bladder over the cut, only to find that the pressure of the sap was enough to extend the bladder. Eventually he devised "aqueo-mercurial gages" (our "mercury gauges") which showed the height of mercury which the sap pressure would raise—a century after the first mercury barometers had been used to measure atmospheric pressure. Hales's ingenuity as an experimenter is very evident in this chapter; he shows it, for example, in his neat method of measuring the pressure with which a root absorbs moisture (Experiment XXI), and in Experiment XXX he very elegantly achieves his result by varying the elements of his experiment one by one. In reading this chapter we have to bear in mind the current hypothesis (suggested at the Royal Society after the reading of Ray's paper) that in plants there was a circulation analogous to that of the blood in animals, and that this circulation took place upwards through the inner part of the stem and downwards through the outermost part. In Experiment XXVI Hales shows that, improbable though it is on the circulation hypothesis, a branch will absorb moisture indifferently in either direction, and similarly in Experiment XXIX he demonstrates that a branch still attached to the tree will actually absorb moisture through its amputated tip, that is, reversing the normal flow up from the roots.

In Chapter III Hales makes a special study of the most striking example of the motion of the sap, that in the vine during the bleeding season. At that period, he shows, even when the vine is without the leaves which normally are so important to the movement of sap, the sap will rise with a pressure equivalent to 36 feet of water. But, on the other hand, the motion of the sap is not due merely to a single pressure from the roots, for gauges attached to different branches of the same tree do not move uniformly.

In the next chapter Hales returns to the circulation hypothesis. By ringing the bark (Plate 13) he shows that the sap does not descend by the most recent growth; on the contrary, it ascends; and (Plates 11, 12) he finds that moisture will enter a tree through the branches or even enter a branch through the leaves, all of which argue against a well-defined circulation comparable to that of the blood in animals.

Earlier, in Experiment XXII and elsewhere, Hales had discovered "many air bubbles issued out of the sap vessels". In the short Chapter V Hales adduces more evidence that air enters into the composition of plants, and this leads him into the extended investigation in Chapter VI (separately titled "Analysis of the Air") in which he applies "chymio-statical experiments" in order to investigate the part "air" plays in the composition of animal, vegetable and mineral substances. Interestingly enough, quantitative methods were for Hales a much more reliable guide in biology than in chemistry; the reduction of chemistry to physics by

Newton and Boyle and later by Hales and various contemporaries had disappointing results. But the chapter as a whole is full of interest. We have to bear in mind that at this period the air was "a fine elastick fluid, with particles of very different natures floating in it"; there were no distinct gases combining to form the air, and in fact air was generally thought to be without chemical properties. It was the isolation and identification of the gases and the recognition of their chemical properties that allowed chemistry to advance so rapidly at the end of the eighteenth century. Hales was not to make this identification, although he was certain that "air" should be adopted "among the chymical principles". His many experiments focused attention on the importance of the air in the composition of substances, and he made extensive investigations into its absorption during combustion, inventing two pieces of apparatus of great significance: a pedestal apparatus (Fig. 35, Plate 16) later used by Priestley, Lavoisier and others, and the famous pneumatic trough (Fig. 38, Plate 17) which may have been for Hales a device essentially for washing his air but which soon provided the necessary technique for collecting gases. In fact, although this long and uneven chapter has provided a problem for historians of science, it seems that through it Hales exerted considerable influence on later chemists, not least on Lavoisier, the central figure in the revolution in chemistry at the end of the century.

In the final chapter, Hales outlines his understanding of the mechanics of plant growth in Newtonian terms, and emphasizes again the rôle of the leaves in passages from which we have already quoted. But the bulk of the chapter is given to a study of the way in which plants (and animals) grow, and Hales shows once again the brilliance of his technique as an experimenter (Plates 18 and 19). In his conclusion, he summarizes the results of his researches (pp. 203–4) and then, as so often before, gives several pages to their practical implications. The mutual benefits of the union between theory and practice had been recognized a century earlier, but examples where theory had in fact assisted practice were still extremely rare. *Vegetable Staticks* was not only a milestone in pure science; it was also one of the early works to illustrate the maxim with which Hales concludes:

> The likeliest method to enable us to make the most judicious observations, and to put us upon the most probable means of improving any art, is to get the best insight we can into the nature and properties of those things which we are desirous to cultivate and improve.

July 3rd, 1961

M. A. HOSKIN

NOTES TO THIS EDITION

The text is that of the 1727 *editio princeps*, except that a few misprints have been corrected where they were likely to cause confusion. Otherwise the original spelling and punctuation has been preserved. Hales supplied two "tables" or indices for the second edition of 1731, and these have been adapted for this edition. Otherwise he made only two notable alterations in 1731, inserting before the last paragraph of Chapter I a long report of the effect on plants and trees of the severe winter of 1728 and repeating with greater care Experiment CXXII. To *Haemastaticks* (1733) he added an Appendix of over a hundred pages containing observations and experiments relating to *Vegetable Staticks*, but these are much inferior in quality to the original material and add little of importance. In fact in the later English editions of the combined two-volumes of *Statical Essays* no attempt was made to incorporate this material into *Vegetable Staticks*.

The standard biography of Hales is *Stephen Hales, D.D., F.R.S.*, by A. E. Clark-Kennedy (Cambridge, 1929, reprinted Ridgewood, N.J., 1965). On the place of Hales in the history of chemistry, see J. R. Partington's *A History of Chemistry*, iii (London, 1962); Eri Yagi, "Stephen Hales' work in chemistry: A Newtonian influence on eighteenth century chemistry", *Japanese Studies in the History of Science*, v (1966) 75–86; and H. Guerlac, "The continental reputation of Stephen Hales", *Archives Internationales d'Histoire des Sciences*, iv (1951) 393–404. On Hales as a Newtonian philosopher, see *Franklin and Newton*, by I. B. Cohen (Philadelphia, 1956, reprinted Cambridge, Mass., 1968). For his work with plants, and in particular for the duplicate "control" system which Hales uses for Experiment CXXII in the second edition, see "Plants and the atmosphere" by Leonard K. Nash, in *Harvard Case Histories in Experimental Science*, edited by J. B. Conant (Cambridge, Mass., 1948).

VEGETABLE STATICKS:

Or, An ACCOUNT of ſome

Statical Experiments

ON THE

SAP in VEGETABLES:

Being an ESSAY towards a

Natural Hiſtory of Vegetation.

Alſo, a SPECIMEN of

An ATTEMPT to Analyſe the AIR,

By a great Variety of

CHYMIO-STATICAL EXPERIMENTS;
Which were read at ſeveral Meetings before
the ROYAL SOCIETY.

Quid eſt in his, in quo non naturæ ratio intelligentis appareat? Tul. de Nat. Deor.
——*Etenim Experimentorum longe major eſt ſubtilitas, quam ſenſûs ipſius*—— *Itaque eo rem deducimus, ut ſenſus tantum de Experimento, Experimentum de re judicet.* Fran. de Verul. Inſtauratio magna.

By *STEPH. HALES*, B. D. F. R. S.
Rector of *Farringdon*, *Hampſhire*, and Miniſter of *Teddington*, *Middleſex*.

LONDON:
Printed for W. and J. INNYS, at the Weſt End of St. *Paul's*; and T. WOODWARD, over-againſt St. *Dunſtan's* Church in *Fleetſtreet*. M, DCC, XXVII.

Feb. 16, 1726–7. *Imprimatur*

ISAAC NEWTON, Pr.Reg.Soc.

TO

His Royal Highness

GEORGE

Prince of Wales.

May it please Your Royal Highness,

I humbly offer the following Experiments to Your Highness's Patronage, to protect them from the reproaches that the ignorant are apt unreasonably to cast on researches of this kind, notwithstanding they are the only solid and rational means whereby we may ever hope to make any real advance in the knowledge of Nature: A knowledge worthy the attainment of Princes.

And as *Solomon*, the greatest and wisest of men, disdained not to inquire into the nature of Plants, *from the Cedar in Lebanon, to the Hyssop that springeth out of the wall:* So it will not, I presume, be an unacceptable entertainment to Your Royal Highness, at least at Your leisure hours; but will rather add to the pleasure, with which vegetable Nature in her prime verdure charms us: To see the steps she takes in her productions, and the wonderful power she therein exerts: The admirable provision she has made for them, not only vigorously to draw to great heights plenty of nourishment from the earth; but also more sublimed and exalted food from the air, that wonderful fluid, which is of such importance to the life of Vegetables and Animals: And which by infinite combinations with natural bodies, produces innumerable surprizing effects; many instances of which I have here shewn.

The searching into the works of Nature, while it delights and inlarges the mind, and strikes us with the strongest assurance of the wisdom and power of the divine Architect, in framing for us so beautiful and well regulated a world, it does at the same time convince us of his constant benevolence and goodness towards us.

That this great Author of Nature may shower down on Your Royal Highness an abundance of his Blessings, both Spiritual and Temporal, is the sincere prayer of

Your Royal Highness's
Most Obedient
Humble Servant,
STEPHEN HALES.

THE PREFACE.

There have been within less than a Century very great and useful discoveries made in the amazingly beautiful structure and nature of the animal œconomy; neither have Plants passed unobserved in this inquisitive age, which has with such diligence extended its inquiries in some degree, into almost every branch of nature's inexhaustible fund of wonderful works.

We find in the Philosophical Transactions, and in the History of the Royal Academy of Sciences, accounts of many curious Experiments and Observations made from time to time on Vegetables, by several ingenious and inquisitive Persons: But our countryman Dr. *Grew* and *Malpighi* were the first, who, tho' in very distant countries did nearly at the same time, unknown to each other, ingage in a very diligent and thorough inquiry into the structure of the vessels of Plants; a province, which till then had layn uncultivated. They have given us very accurate and faithful accounts of the structure of the parts, which they carefully traced, from the first minute origin, the seminal Plants, to their full growth and maturity, thro' their Roots, Trunk, Bark, Branches, Gems, Shoots, Leaves, Blossoms and Fruit. In all which they observed an exact and regular symetry of Parts most curiously wrought in such manner, that the great work of Vegetation might effectually be carried on, by the uniform co-operation of the several Parts, according to the different offices assigned them by nature.

Had they fortuned to have fallen into this statical way of inquiry, persons of their great application and sagacity had doubtless made considerable advances in the knowledge of the several quantities of nourishment, which Plants imbibe and perspire, and thereby to see what influence the different states of Air have on them. This is the likeliest method to find out the Sap's velocity, and the force with which it is imbibed: As also to estimate the great power that nature exerts in extending and pushing forth her productions, by the expansion of the Sap.

About 20 years since, I made several haemastatical Experiments on Dogs, and 6 Years afterwards repeated the same on Horses and other Animals, in order to find out the real force of the blood in the Arteries, some of which are mentioned in the third chapter of this book: At which times I wished I could have made the like Experiments, to discover the force of the Sap in Vegetables; but despaired of ever effecting it, till about seven years since, by mere accident I hit upon it, while I was endeavouring by several ways to stop the bleeding of an old stem of a Vine, which was cut too near the bleeding season, which I feared might kill it: Having, after other means proved ineffectual, tyed a piece of bladder over the transverse cut of the Stem, I found the force of the Sap did greatly extend the bladder; whence I concluded, that if a long glass tube were fixed there in the same manner, as I had before done to the Arteries of several living Animals, I should thereby obtain the real ascending force of the Sap in that Stem, which succeeded according to my expectation, and hence it is, that I have been insensibly led on, to make farther and farther researches by variety of Experiments.

As the art of Physick has of late years been much improved by a greater knowledge of the animal œconomy; so doubtless a farther insight into the vegetable œconomy must needs proportionably improve our skill in Agriculture and Gardening, which gives me reason to hope, that inquiries of this kind will be acceptable to many, who are intent upon improving those innocent, delightful, and beneficial Arts: Since they cannot be insensible that the most rational ground for Success in this laudable pursuit must arise from a greater insight into the nature of Plants.

Finding by many Experiments in the fifth chapter, that the Air is plentifully inspired by Vegetables, not only at their roots, but also thro' several parts of their Trunks and Branches; this put me upon making a more particular inquiry into the nature of the Air; and to discover, if possible, wherein its great importance to the life and support of Vegetables might consist; on which account I was obliged to delay the publication of the rest of these Experiments, which were read two years since before the Royal Society, till I had made some progress in this inquiry. An account of which I have given in the sixth chapter.

Where it appears by many chymio-statical Experiments, that there is diffused thro' all natural, mutually attracting bodies, a large proportion of particles, which, as the first great Author of this important discovery, Sir *Isaac Newton*, observes, are capable of being thrown off from dense bodies by heat or fermentation into a vigorously elastick and permanently repelling state: And also of returning by fermentation, and sometimes without it, into dense bodies: It is by this amphibious property of the air, that the main and principal operations of Nature are carried on; for a mass of mutually attracting particles, without being blended with a due proportion of elastick repelling ones, would in many cases soon coalesce into a sluggish lump. It is by these properties of the particles of matter that he solves the principal Phœnomena of Nature. And Dr. *Freind* has from the same principles given a very ingenious *Rationale* of the chief operations in Chymistry. It is therefore of importance to have these very operative properties of natural bodies further ascertained by more Experiments and Observations: And it is with satisfaction that we see them more and more confirmed to us, by every farther enquiry we make; as the following Experiments will plainly prove, by shewing how great the power of the attraction of acid sulphureous particles must be at some little distance from the point of contact, to be able most readily to subdue and fix elastick aereal particles, which repel with a force superior to vast incumbent pressures: Which particles we find are thereby changed from a strongly repelling, to as strongly an attracting state: And that elasticity is no immutable property of air, is further evident from these Experiments; because it were impossible for such great quantities of it to be confined in the substances of Animals and Vegetables, in an elastick state, without rending their constituent parts with a vast explosion.

I have been careful in making, and faithful in relating the result of these Experiments, and wish I could be as happy in drawing the proper inferences from them. However I may fall short at first setting out in this statical way of inquiring into the nature of Plants, yet there is good reason to believe that considerable advances in the knowledge of their nature may in process of time be made, by researches of this kind.

And I hope the publication of this Specimen of what I have hitherto done, will put others upon the same pursuits, there being in so large a field, and among such an innumerable variety of subjects, abundant room for many heads and hands to be employed in the work: For the wonderful and secret operations of Nature are so involved and intricate, so far out of the reach of our senses, as they present themselves to us in their natural order, that it is impossible for the most sagacious and penetrating genius to pry into them, unless he will be at the pains of analysing Nature, by a numerous and regular series of Experiments; which are the only solid foundation whence we may reasonably expect to make any advance, in the real knowledge of the nature of things.

I must not omit here publickly to acknowledge, that I have in several respects been much obliged to my ingenious and learned neighbour and friend *Robert Mathers* of the *Middle Temple*, Esq; for his assistance herein.

The Contents.

Chap. I.

Chap. II.

Chap. III.

Chap. IV.

Chap. V.

Chap. VI.

Chap. VII.

The Introduction.

The farther researches we make into this admirable scene of things, the more beauty and harmony we see in them: And the stronger and clearer convictions they give us, of the being, power and wisdom of the divine Architect, who has made all things to concur with a wonderful conformity, in carrying on, by various and innumerable combinations of matter, such a circulation of causes, and effects, as was necessary to the great ends of nature.

And since we are assured that the all-wise Creator has observed the most exact proportions, *of number, weight and measure*, in the make of all things; the most likely way therefore, to get any insight into the nature of those parts of the creation, which come within our observation, must in all reason be to number, weigh and measure. And we have much encouragement to pursue this method, of searching into the nature of things, from the great success that has attended any attempts of this kind.

Thus, in relation to those Planets which revolve about our Sun, the great Philosopher of our age has, by numbering and measuring, discovered the exact proportions that are observed in their periodical revolutions and distances from their common centers of motion and gravity: And that God has not only *comprehended the dust of the earth in a measure, and weighed the mountains in scales, and the hills in a balance*, Isai. xl. 12. but that he also holds the vast revolving Globes, of this our solar System, most exactly poised on their common center of gravity.

And if we reflect upon the discoveries that have been made in the animal œconomy, we shall find that the most considerable and rational accounts of it have been chiefly owing to the statical examination of their fluids, *viz.* by enquiring what quantity of fluids, and solids dissolved into fluids, the animal daily takes in for its support and nourishment: And with what force and different rapidities those fluids are carried about in their proper channels, according to the different secretions that are to be made from them: And in what proportion the recrementitious fluid is conveyed away, to make room for fresh supplies; and what portion of this recrement nature allots to be carried off, by the several kinds of emunctories and excretory ducts.

And since in vegetables, their growth and the preservation of their vegetable life is promoted and maintained, as in animals, by the very plentiful and regular motion of their fluids, which are the vehicles ordained by nature, to carry proper nutriment to every part; it is therefore reasonable to hope, that in them also, by the same method of inquiry, considerable discoveries may in time be made, there being, in many respects, a great analogy between plants and animals.

CHAP. I.

Experiments, shewing the quantities imbibed and perspired by Plants and Trees.

EXPERIMENT I.

July 3. 1724. in order to find out the quantity imbibed and perspired by the Sun-Flower, I took a garden-pot (Fig. 1.) with a large *Sun-Flower, a,* 3 feet+½ high, which was purposely planted in it when young.

I covered the pot with a plate of thin milled lead, and cemented all the joints fast, so as no vapour could pass, but only air, thro' a small glass tube *d* nine inches long, which was fixed purposely near the stem of the plant, to make a free communication with the outward air, and that under the leaden plate.

I cemented also another short glass tube *g* into the plate, two inches long and one inch in diameter. Thro' this tube I watered the plant, and then stopped it up with a cork; I stopped up also the holes *i, l* at the bottom of the pot with corks.

I weighed this pot and plant morning and evening, for fifteen several days, from *July* 3. to *Aug*. 8. after which I cut off the plant close to the leaden plate, and then covered the stump well with cement; and upon weighing found there perspired thro' the unglazed porous pot two ounces every twelve hours day, which being allowed in the daily weighing of the plant and pot, I found the greatest perspiration of twelve hours in a very warm dry day, to be one pound fourteen ounces; the middle rate of perspiration one pound four ounces. The perspiration of a dry warm night, without any sensible dew, was about three ounces; but when any sensible, tho' small dew, then the perspiration was nothing; and when a large dew, or some little rain in the night, the plant and pot was increased in weight two or three ounces. N. B. *The weights I made use of were* Avoirdupoise *weights*.

I cut off all the leaves of this plant, and laid them in five several parcels, according to their several sizes, and then measured the surface of a leaf of each parcel, by laying over it a large lattice made with threads, in which the little squares were $\frac{1}{4}$ of an inch each; by numbering of which I had the surface of the leaves in square inches, which multiplied by the number of the leaves in the corresponding parcels, gave me the area of all the leaves; by which means I found the surface of the whole plant, above ground, to be equal to 5616 square inches, or 39 square feet.

I dug up another Sun-flower, nearly of the same size, which had eight main roots, reaching fifteen inches deep and sideways from the stem: It had besides a very thick bush of lateral roots, from the eight main roots, which extended every way in a Hemisphere, about nine inches from the stem and main roots.

In order to get an estimate of the length of all the roots, I took one of the main roots, with its laterals, and measured and weighed them, and then weighed the other seven roots, with their laterals, by which means I found the sum of the length of all the roots to be no less than 1448 feet.

And supposing the periphery of these roots at a medium, to be $\frac{10}{76}$ of an inch, then their surface will be 2286 square inches, or 15·8 square feet; that is, equal to $\frac{3}{8}$ of the surface of the plant above ground.

If, as above, twenty ounces of water, at a medium, perspired in twelve hours day (*i.e.*) thirty four cubick inches of water (a cubick inch of water weighing 254 grains) then the thirty four cubick inches divided by the surface of all the roots, is$=$2286 square inches; (*i.e.*) $\frac{34}{2286}$ is$=\frac{1}{67}$ this gives the depth of water imbibed by the whole surface of the roots *viz.* $\frac{1}{67}$ part of an inch.

And the surface of the plant above ground being 5616 square inches, by which dividing the 34 cubick inches, *viz.* $\frac{34}{5616}=\frac{1}{165}$, this gives the depth perspired by the whole surface of the plant above ground, *viz.* $\frac{1}{165}$ part of an inch.

Hence, the velocity with which water enters the surface of the roots to supply the expence of perspiration, is to the velocity, with which the sap perspires, as 165 : 67, or as $\frac{1}{67}:\frac{1}{165}$, or as nearly as 5 : 2.

The area of the transverse cut of the middle of the stem is a square inch; therefore the areas, on the surface of the leaves, the roots, and stem, are 5616, 2286, 1.

The velocities, in the surface of the leaves, roots, and transverse cut of the stem, are gained by a reciprocal proportion of the surfaces.

Area of		velocity		or as	
	leaves $=5616$		$=\frac{1}{5616}$		$\frac{1}{165}$ inch
	roots $=2286$		$=\frac{1}{2286}$		$\frac{1}{67}$ inch
	stem $=$ 1		$=$ 1		34 inch.

Now, their perspiring 34 cubick inches in twelve hours day, there must so much pass thro' the stem in that time; and the velocity would be at the rate of 34 inches in twelve hours, if the stem were quite hollow.

In order therefore to find out the quantity of solid matter in the stem, *July 27th* at 7 *a.m.* I cut up even with the ground a Sunflower; it weighed 3 pounds; in thirty days it was very dry, and had wasted in all 2 pounds 4 ounces; that is $\frac{3}{4}$ of its whole weight: So here is a fourth part left for solid parts in the stem, (by throwing a piece of green Sun-flower stem into water, I found it very near of the same specifick gravity with water) which filling up so much of the stem, the velocity of the sap must be increased proportionably, *viz.* $\frac{1}{3}$ part more, (by reason of the reciprocal proportion) that 34 cubick inches may pass the stem in twelve hours; whence its velocity in the stem will be $45\frac{1}{3}$ inches in twelve hours, supposing there be no circulation nor return of the sap downwards.

If there be added to 34, (which is the least velocity) $\frac{1}{3}$ of it$=11\frac{1}{3}$ this gives the greatest velocity, *viz.* $45\frac{1}{3}$ The spaces being as 3 : 4. the velocities will be 4 : 3 :: $45\frac{1}{3}$: 34.

But if we suppose the pores in the surface of the leaves to bear the same proportion, as the area of the sap vessels in the stem do to the area of the stem; then the velocity, both in the leaves, root and stem, will be increased in the same proportion.

A pretty exact account having been taken, of the weight, size, and surface of this plant, and of the quantities it has imbibed and perspired, it may not be improper here, to enter into a comparison, of what is taken in and perspired by a human body, and this plant.

The weight of a well sized man is equal to 160 pound: The weight of the Sun-flower is three pounds, so their weights are to each other as 160 : 3, or as 53 : 1.

The surface of such human body is 15 square feet, or 2160 square inches.

The surface of the Sun-flower is 5616 square inches, so its surface is to the surface of a human body as 26 : 10.

The quantity perspired by a man in twenty four hours is 31 ounces, as Dr. *Keill* found, *vide Medicina Statica Britannica*, p. 14.

The quantity perspired by the plant, in the same time, is 22 ounces, allowing two ounces for the perspiration of the beginning, and ending of the night in *July*, viz. after evening and before morning weighing, just before and after night.

So the perspiration of a man to the Sun-flower is as 141 : 100.

Abating the six ounces of the thirty one ounces, to be carried off by respiration from the lungs in the twenty four hours; (which I have found by certain experiment to be so much if not more) the twenty five ounces multiplied by $437+\frac{1}{2}$, the number of grains in an ounce *Avoirdupois*, the product is $10937\frac{1}{2}$ grains; which divided by 254, the number of grains in a cubick inch of water gives 43 cubick inches perspired by a man: Which divided by the surface of his body, *viz.* 2160 square inches the quotient is $\frac{1}{50}$ part of a cubick inch perspired off a square inch in twenty four hours. Therefore in equal surfaces and equal times the man perspires $\frac{1}{50}$, the plant $\frac{1}{165}$, or as 50 : 15.

Which excess in the man is occasioned by the very different degrees of heat in each: For the heat of the plant cannot be greater than the heat of the circum-ambient air, which heat in summer is from 25 to 35 degrees above the freezing point, (*vide Exp.* 20.) but the heat of the warmest external parts of a man's body is 54 such degrees; and the heat of the blood 64 degrees; which is nearly equal to water heated to such a degree, as a man can well bear to hold his hand in, stirring it about; which heat is sufficient to make a plentiful evaporation.

Qu. Since then the perspirations of equal areas in a man and a Sunflower, are to each other as 165 : 50. or as $3\frac{1}{3}$: 1; and since the degrees of heat are as 2 : 1, must not the sum or quantity of the areas of the pores lying in equal surfaces, in the man and Sunflower, be as 16 : 1? for it seems that the quantities of the evaporated fluid will be as the degrees of heat, and the sum of the areas of the pores taken together.

Dr. *Keill*, by estimating the quantities of the several evacuations of his body, found that he eat and drank, every 24 hours, 4 pounds 10 ounces.

The Sunflower imbibed and perspired in the same time twenty two ounces, so the man's food, to that of the plant, is as 74 ounces to 22 ounces, or as 7 : 2.

But compared bulk for bulk, the plant imbibes seventeen times more fresh food than the man: For deducting five ounces, which Dr. *Keill* allows for the *fæces alvi,* there will remain sixty nine ounces of fresh liquor, which enters a man's veins; and an equal quantity passes off every 24 hours. Then it will be found, that seventeen times more new fluid enters the sap vessels of the plant, and passes off in 24 hours, than there enters the veins of a man, and passes off in the same time.

And since, compared bulk for bulk, the plant perspires seventeen times more than the man, it was therefore very necessary, by giving it an extensive surface, to make a large provision for a plentiful perspiration in the plant, which has no other way of discharging superfluities; whereas there is provision made in man, to carry off above half of what he takes in, by other evacuations.

For since neither the surface of his body was extensive enough to cause sufficient exhalation, nor the additional wreak, arising from the heat of his blood, could carry off above half the fluid, which was necessary to be discharged every twenty four hours; there was a necessity of providing the kidneys, to percolate the other half thro'.

And whereas it is found, that seventeen times more enters, bulk for bulk, into the sap vessels of the plant, than into the veins of a man, and goes off in twenty four hours: One reason of this greater plenty of fresh fluid, in the vegetable than the animal body, may be, because the fluid which is filtrated thro' the roots immediately from the earth, is not near so full fraighted with nutritive particles as the chyle which enters the lacteals of animals; which defect it was necessary to supply by the entrance of a much greater quantity of fluid.

And the motion of the sap is thereby much accelerated, which in the heartless vegetable would otherwise be very slow; it having probably only a progressive and not a circulating motion, as in animals.

Since then a plentiful perspiration is found so necessary for the health of a plant or tree, 'tis probable that many of their distempers are owing to a stoppage of this perspiration, by inclement air.

The perspiration in men is often stopped to a fatal degree; not only by the inclemency of the air, but by intemperance, and violent heats and colds. But the more temperate vegetable's perspiration can be stopped only by inclement air; unless by an unkindly soil, or want of genial moisture it is deprived of proper or sufficient nourishment.

As Dr. *Keill* observed in himself a considerable latitude of degrees of healthy perspiration, from a pound and a half to 3 pounds; I have also observed, a healthy latitude of perspiration in this Sunflower, from sixteen to twenty eight ounces in twelve hours day. The more it was watered, the more plentifully it perspired, (*cæteris paribus*) and with scanty watering the perspiration much abated.

EXPERIMENT II.

From *July* 3*d.* to *Aug.* 3*d.* I weighed for nine several mornings and evenings a middle sized *Cabbage plant*, which grew in a garden pot, and was prepared with a leaden cover as the Sunflower, *Exper.* 1*st.* Its greatest perspiration in 12 hours day was 1 pound 9 ounces; its middle perspiration 1 pound 3 ounces, $=$ 32 cubick inches. Its surface 2736 square inches, or 19 square feet. Whence dividing the 32 cubick inches by 2736 square inches, it will be found that a little more than the $\frac{1}{86}$ of an inch depth perspires off its surface in 12 hours day.

The area of the middle of the Cabbage stem is $\frac{100}{156}$ of a square inch; hence the velocity of the sap in the stem, is to the velocity of the perspiring sap, on the surface of the leaves, as $2736 : \frac{100}{156} :: 4268 : 1$. for $\frac{2736 \times 156}{100} = 4268$. But if an allowance is to be made for the solid parts of the stem (by which the passage is narrowed) the velocity will be proportionably increased.

The length of all its roots 470 feet, their periphery at a medium $\frac{1}{22}$ of an inch, hence their area will be 256 square inches nearly; which being so small, in proportion to the area of the leaves, the sap must go with near eleven times the velocity through the surface of the roots, that it does thro' the surface of the leaves.

And setting the roots at a medium at 12 inches long, they must occupy a hemisphere of earth two feet diameter, that is 2. 1 cubick feet of earth.

By comparing the surfaces of the roots of plants, with the surface of the same plant above ground, we see the necessity of cutting off many branches, from a transplanted tree: for if 256 square inches of root in surface was necessary to maintain this Cabbage in a healthy natural state: suppose upon digging it up, in order to transplant, half the roots be cut off (which is the case of most young transplanted trees) then it's plain, that but half the usual nourishment can be carried up, through the roots, on that account; and a very much less proportion on account of the small hemisphere of earth, the new planted shortened roots occupy; and on account of the loose position of the new turned earth, which touches the roots at first but in few points. This (as well as experience) strongly evinces the great necessity of well watering new plantations.

Which yet must be done with caution, for the skilful and ingenious Mr. *Philip Miller*, Gardiner of the Botanick garden at *Chelsea*, in his very useful Gardiners and Florists Dictionary, says, "That he has often seen trees, that have had too much water given them after planting, which has rotted all the young fibres, as fast as they have been pushed out; and so many times has killed the tree." *Supplement* Vol. II. *of planting*. And I observed that the dwarf pear-tree, whose root was set in water, in *Exper*. 7. decreased very much daily in the quantity imbibed; *viz*. because the sap vessels of the roots, like those of the cut off boughs, in the same experiment, were so saturated and clogged with moisture, by standing in water, that more of it could not be drawn up to support the leaves.

Experiment III.

From *July* 28. to *Aug.* 25. I weighed for twelve several mornings and evenings, a thriving *Vine* growing in a pot; I was furnished, with this and other trees, from his Majesty's garden at *Hampton-court*, by the favour of the eminent Mr. *Wise*. This vine was prepared with a cover, as the Sun-flower was. Its greatest perspiration in 12 hours day, was 6 ounces + 240 grains; its middle perspiration 5 ounces + 240 grains = to $9\frac{1}{2}$ cubick inches.

The surface of its leaves was 1820 square inches, or 12 square feet + 92 square inches; whence dividing $9\frac{1}{2}$ cubick inches, by the area of the leaves, it is found that $\frac{1}{191}$ part of an inch depth, perspires off in 12 hours day.

The area of a transverse cut of its stem, was equal to $\frac{1}{4}$ of a square inch: hence the sap's velocity here to its velocity on the surface of the leaves, will be as $1820 \times 4 = 7280 : 1$. Then the real velocity of the sap's motion in the stem is $= \frac{7280}{191} = 38$ inches in twelve hours.

This is supposing the stem to be a hollow tube: but by drying a large vine branch (in the chimney corner) which I cut off, in the bleeding season, I found the solid parts were $\frac{3}{4}$ of the stem; hence the cavity thro' which the sap passes, being so much narrowed, its velocity will be 4 times as great, *viz.* 152 inches in 12 hours.

But it is further to be considered, that if the sap moves in the form of vapor and not of water, being thereby rarified, its velocity will be increased in a direct proportion of the spaces, which the same quantity of water and vapor would occupy: And if the vapor is supposed to occupy 10 times the space which it did, when in the form of water, then it must move 10 times faster; so that the same quantity or weight of each may pass in the same time, thro' the same bore or tube: And such allowance ought to be made in all these calculations concerning the motion of the sap in vegetables.

Experiment IV.

From *July* 29. to *Aug.* 25. I weighed for 12 several mornings and evenings, a *paradise stock Apple-tree*, which grew in a garden pot, covered with lead, as the Sun-flower: it had not a bushy head full of leaves, but thin spread, being in all but 163 leaves; whose surface was equal to 1589 square inches, or 11 square feet + 5 square inches.

The greatest quantity it perspired in 12 hours day, was 11 ounces, its middle quantity 9 ounces, or $15\frac{1}{2}$ cubick inches.

The $15\frac{1}{2}$ cubick inches perspired, divided by the surface 1589 square inches, gives the depth perspired off the surface in 12 hours day, *viz.* $\frac{1}{104}$ of an inch.

The area of a transverse cut of its stem, $\frac{1}{4}$ of an inch square, whence the sap's velocity here, will be to its velocity on the surface of the leaves as $1589 \times 4 = 6356 : 1$.

Experiment V.

From *July* 28. to *Aug.* 25. I weighed for 10 several mornings and evenings a very thriving *Limon-tree*, which grew in a garden pot, and was covered as above: Its greatest perspiration in 12 hours day was 8 ounces, its middle perspiration 6 ounces, equal to $10\frac{1}{3}$ cubick inches. In the night it perspired sometimes half an ounce, sometimes nothing, and sometimes increased 1 or 2 ounces in weight, by large dew or rain.

The surface of its leaves was 2557 square inches, or 17 square feet + 59 square inches; dividing then the $10\frac{1}{2}$ cubick inches perspired by this surface, gives the depth perspired in 12 hours day, *viz.* $\frac{1}{243}$ of an inch.

So the several foregoing perspirations in equal areas are,

- $\frac{1}{191}$ in the vine in 12 hours day.
- $\frac{1}{50}$ in a man, in a day and a night.
- $\frac{1}{165}$ in a Sunflower, in a day and night.
- $\frac{1}{86}$ in a cabbage, in 12 hours day.
- $\frac{1}{104}$ in an apple-tree, in 12 hours day
- $\frac{1}{243}$ in a limon-tree, in 12 hours day.

The area of the transverse cut of the stem of this Limon-tree was $=1.44$ of a square inch; hence the sap's velocity here, will be to its velocity on the surface of the leaves, as $1768:1$ for $\frac{2557\times 100}{144}=1768$. This is supposing the whole stem to be a hollow tube; but the velocity will be increased both in the stem and the leaves, in proportion as the passage of the sap is narrowed by the solid parts.

By comparing the very different degrees of perspiration, in these 5 plants and trees, we may observe, that the limon-tree, which is an ever-green, perspires much less than the Sunflower, or than the Vine or the Apple-tree, whose leaves fall off in the winter; and as they perspire less, so are they the better able to survive the winter's cold, because they want proportionably but a very small supply of fresh nourishment to support them: like the exangueous tribe of animals, frogs, toads, tortoises, serpents, insects, *&c.* which as they perspire little; so do they live the whole winter without food. And this I find holds true in 12 other different sorts of ever-greens, on which I have made Experiments.

The above mentioned Mr. *Miller* made the like experiments in the Botanick-garden at *Chelsea*, on a plantain-tree, an aloe, and a paradise apple-tree; which he weighed morning, noon, and night, for several successive days. I shall here insert the diaries of them, as he communicated them to me, that the influence of the different temperatures of the air, on the perspiration of these plants, may the better be seen.

The pots which he made use of were glazed, and had no holes in their bottoms, as garden pots usually have; so that all the moisture, which was wanting in them upon weighing, must necessarily be imbibed, by the roots of those plants, and thence perspired off thro' their leaves.

A diary of the perspiration of the Musa Arbor, *or* Plantain-tree *of the* West-Indies. *The whole surface of the plant was* 14 *square feet,* 8+½ *inches. The different degrees of heat of the air, are here noted by the degrees above the freezing point in my* Thermom. *describ'd in* Exp. 20.

1726 May	Weight at 6 Morn. pd. ou.	Therm.	Weight at 12 Noon. pd. ou.	Therm.	Weight at 6 Even. pd. ou.	Therm.	*N.B.* This plant stood in a stove, with a small fire in it; the aspect of the stove was South-east.
17	38 5	31	38 0	38	37 14	34	
18	37 15	29	37 5½	45	37 3½	31	A hot clear day. This morning he observed large drops of water at the extremity of every leaf, and we may observe that it perspires very much this day.
19	37 4	32	37 2	35	37 0	31	An extream hot clear day.
20	36 14	34	36 12	48	36 11	36	Moderately hot but clear.
21	36 10	30	37 0	50	36 15	44	This morn, 12 ounces of water poured into the pot. Mixture of Sun and Clouds.
22	36 14	31			36 11½	35	Much thunder, some rain and hail at a distance.
23	36 6	32	36 5½	32½	36 5	31	A gloomy day but no rain.

This evening 12 ounces of water were poured into the pot; and it was removed from the stove into a cool room, where it had a free air but no Sun, the windows being North-west.

1726 May	Weight at 6 Morn. pd. ou.	Therm.	Weight at 12 Noon. pd. ou.	Therm.	Weight at 6 Even. pd. ou.	Therm.	
24	37 00	27	37 00	27½	36 15½	25½	Calm cloudy weather.
25	37 00	21½	36 14½	26	36 13	23	A pretty clear day.
26	36 12	22	36 11	25	36 10	24	A hot day.
27	36 10½	23	36 6¾	26½	36 6	25½	A very hot day.
28	36 6	22½	36 5	24	36 3½	23	Some rain and cloudy. At this time, the under leaves of the plant began to wither and decay; and the top leaf to unfold and spread abroad; but they are observed never to grow bigger, after they are fully opened.
29	36 2	20	36 2½	21½	36 1	22	A temperate day.
30	36 1½	19	36 1	21	36 0	19	Temperate weather not very clear.
June 1	35 15	18	35 14½	19½	35 13½	18	Some rain. The whole plant begins to change colour, and appear sickly.
2	35 12	19½	35 11½	23	35 11	21½	He then removed the plant into the stove again in order to recover it; but it continued to fade, and in two or three days dyed.
3	35 10	28½	35 4	36	35 1½	34	A cool and cloudy day.
4	35 00	26	34 14	31	34 11	29	A warm day; and the whole plant decayed.

We may observe from this diary, that this plant, when in the stove, usually perspired more in 6 hours before noon than in 6 hours after noon; and that it perspired much less in the night than in the day time: And sometimes increased in weight in the night, by imbibing the moisture of the ambient air; and that both in the stove and in the cool room. Upon making an estimate of the quantity perspired off a square inch of this plant, in 12 hours day, it comes but to $\frac{1}{112}$ of a cubick inch; on the 18*th* day of *May*, when by far its greatest perspiration was; for on several other days it was much less.

A diary of the Aloe Africana Caulescens *foliis spinosis, maculis ab utraque parte Albicantibus notatis, Commelini hort. Amst. commonly called the* Carolina Aloe. *It was a large plant of its kind. It stood in a glass-case, which had a South aspect without a fire.*

1726 May	Weight at 6 Morn. pd. ou.	Therm.	Weight at 12 Noon	Therm.	Weight at 6 Night.	Therm.	
18	41 6	35	41 2½	36	41 3	30½	
19	41 1½	28½	40 14	31½	40 12	30	
20	40 12½	26½	40 10	31	40 8½	29½	
21	40 9½	27	40 6¾	30	40 5½	28	
22	40 6	25½	40 5½	29	40 4	27½	This evening promising some rain, he set the pot out, to receive a little, and then wiping the leaden surface of the pot dry, he set it into the glass-case again.
23	41 10	24½	41 6½	29	41 5	27½	Now the pot broke, and hindered any further observations.

We may observe, that this Aloe increased in weight most nights, and perspired most in the morning.

A diary of a small Paradise-Apple, *with one upright stem 4 feet high; and two small lateral branches about 8 inches long. This plant stood under a cover of wood which was open on all sides.*

1726 May							
18	37 4	1	37 3	22	37 1	20	
19	37 1	17½	36 14	21	36 13½	19	
20	36 12	18½	36 10½	23	36 9	20½	The leaves very dry,
21	36 7	17	36 5	21½	36 4	20	and become speckled for want of dew.
22	36 3½	18½	36 1	24	36 2½	22½	Then he removed the plant into the stove, to try what effect that would have on its perspiration.

1726 May							
24	36 00	26	35 8	$37\frac{1}{2}$	35 $5\frac{1}{2}$	$34\frac{1}{2}$	At this time the leaves were withered with the heat and hung down as if they would fall off.
25	35 4	$32\frac{1}{2}$	35 1	36	35 00	30	At this time several of the leaves began to fall off.
26	34 9	$28\frac{1}{2}$	34 $6\frac{1}{2}$	34	34 1	32	All the leaves fallen off, except a few small ones, at the extremities of the branches which had put out, since the plant was in the stove.
27	33 $7\frac{1}{2}$	28					The earth it stood in was very moist all the time.

In *October* 1725. Mr. *Miller* took up an *African Briony-root*, which when cleared from the mould weighed eight ounces $\frac{1}{2}$; he laid it on a shelf in the stove, where it remained till the *March* following; when upon weighing he found it had lost of its weight. In *April* it shot out 4 branches, two of which were $3\frac{1}{2}$ feet long, the other two were one of them 14 inches, the other 9 inches, in length: These all produced fair large leaves, it had lost $1\frac{3}{4}$ ounces in weight, and in three weeks more it lost $2\frac{1}{4}$ ounces more, and was much withered.

Experiment VI.

Spear-mint being a plant that thrives most kindly in water, (in order the more accurately to observe what water it would imbibe, and perspire by night and day, in wet or dry weather) I cemented at *r* a plant of it *m*, into the inverted syphon *r y x b* (Fig. 2.) the syphon was $\frac{1}{4}$ inch diam. at *b*, but larger at *r*.

I filled it full of water, the plant imbibed the water so as to make it fall in the day, (in *March*) near an inch and half from *b* to *t*, and in the night $\frac{1}{4}$ inch from *t* to *i*: but one night, when it was so cold, as to make the *Thermometer* sink to the freezing point, then the mint imbibed nothing, but hung down its head; as did also the young beans in the garden, their sap being greatly condensed by cold. In a rainy day the mint imbibed very little.

I pursued this Experiment no farther, Dr. *Woodward* having long since, from several curious experiments and observations, given an account in the Philosophical Transactions, of the plentiful perspirations of this plant.

Experiment VII.

In *August*, I dug up a large dwarf *Pear-tree*, which weighed 71 pounds 8 ounces; I set its root in a known quantity of water; it imbibed 15 pounds of water in 10 hours day, and perspired at the same time 15 pounds 8 ounces.

In *July* and *August* I cut off several branches of Apple-trees, Pear, Cherry, and Apricot-trees, two of a sort; they were of several sizes from 3 to 6 feet long, with proportional lateral branches; and the transverse cut of the largest part of their stems was about an inch diameter.

I stripped the leaves off of one bough of each sort, and then set their stems in separate glasses, pouring in known quantities of water.

The boughs with leaves on them imbibed some 15 ounces, some 20 ounces, 25 or 30 ounces in 12 hours day, more or less in proportion to the quantity of leaves they had; and when I weighed them at night they were lighter than in the morning.

While those without leaves imbibed but one ounce, and were heavier in the evening than in the morning, they having perspired little.

The quantity imbibed by those with leaves decreased very much every day, the sap vessels being probably shrunk, at the transverse cut, and too much saturate with water, to let any more pass; so that usually in 4 to 5 days the leaves faded and withered much.

I repeated the same experiment with Elm-branches, Oak, Osier, Willow, Sallow, Aspen, Curran, Goosberry; and Philbert branches; but none of these imbibed so much as the foregoing, and several sorts of ever-greens very much less.

Experiment VIII.

August 15. I cut off a large *Russet-pippin*, with two inches stem, and its 12 adjoining leaves; I set the stem in a little viol of water it imbibed and perspired in three days $\frac{4}{5}$ of an ounce.

At the same time I cut off from the same tree another bearing twig of the same length, with 12 leaves on it, but no apple; it imbibed in the same three days near $\frac{3}{4}$ of an ounce.

About the same time I set in a viol of water a short stem of the same tree, with two large apples on it without leaves; they imbibed near $\frac{1}{4}$ ounce in two days.

So in this Experiment, the apple and the leaves imbibe $\frac{4}{5}$ of an ounce; the leaves alone near $\frac{3}{4}$ but the two large apples imbibed and perspired but $\frac{1}{3}$ part so much as the 12 leaves; then one apple imbibed the $\frac{1}{6}$ part of what was imbibed by the 12 leaves, therefore two leaves imbibe and perspire as much as one apple; whence their perspirations seem to be proportiònable to their surfaces; the surface of the apple being nearly equal to the sum of the upper and under surfaces of the two leaves.

Whence it is probable, that the use of these leaves, (which are placed, just where the fruit joins to the tree) is to bring nourishment to the fruit. And accordingly I observe that the leaves, next adjoining to blossoms, are, in the spring, very much expanded, when the other leaves, on barren shoots, are but beginning to shoot: And that all peach leaves are pretty large before the blossom goes off: And that in apples and pears the leaves are one third or half grown, before the blossom blows: So provident is nature in making timely provision for the nourishing the yet embrio fruit.

Experiment IX.

July 15. I cut off two thriving *Hop-vines* near the ground, in a thick shady part of the garden, the pole still standing; I striped the leaves off one of these vines, and set both their stems, in known quantities of water, in little bottles; that with leaves imbibed in 12 hours day 4 ounces, and that without leaves $\frac{3}{4}$ of an ounce.

I took another hop pole with its vines on it, and carried it out of the hop ground, into a free open exposure; these imbibed and perspired as much more as the former in the hop-ground: Which is doubtless the reason why the hop-vines on the outsides of gardens, where most exposed to the air, are short and poor, in comparison of those in the middle of the ground; *viz.* because being much dried, their fibres harden sooner, and therefore they cannot grow so kindly as those in the middle of the ground; which by shade are always kept moister, and more ductile.

Now there being 1000 hills in an acre of hop-ground, and each hill having three poles, and each pole three vines, the number of vines will be 9000; each of which imbibing 4 ounces, the sum of all the ounces, imbibed in an acre in 12 hours day, will be 36000 ounces, = 15750000 grains = 62007 cubick inches or 220 gallons; which divided by 6272640, the number of square inches in an acre, it will be found, that the quantity of liquor perspired by all the hop-vines, will be equal to an area of liquor, as broad as an acre, and $\frac{1}{101}$ part of an inch deep, besides what evaporated from the earth.

And this quantity of moisture in a kindly state of the air is daily carried off, in a sufficient quantity, to keep the hops in a healthy state; but in a rainy moist state of air, without a due mixture of dry weather, too much moisture hovers about the hops, so as to hinder in a good measure the kindly perspiration of the leaves, whereby the stagnating sap corrupts, and breeds moldy fen, which often spoils vast quantities of flourishing hop-grounds. This was the case in the year 1723, when 10 or 14 days almost continual rains fell, about the latter half of *July*, after 4 months dry weather; upon which the most flourishing and promising hops were all infected with mold or fen, in their leaves and fruit, while the then poor and unpromising hops escaped, and produced plenty; because they being small, did not perspire so great a quantity as the others; nor did they confine the perspired vapor, so much as the large thriving vines did, in their shady thickets.

This rain on the then warm earth made the grass shoot out, as fast as if it were in a hot-bed; and the apples grew so precipitately, that they were of a very flashy constitution, so as to rot more remarkably than had ever been remembred.

The planters observe, that when a mold or fen has once seized any part of the ground, it soon runs over the whole; and that the grass, and other herbs, under the hops, are infected with it.

Probably because the small seeds of this quick growing mold, which soon come to maturity, are blown over the whole ground: Which spreading of the seed may be the reason why some grounds are infected with fen for several years successively; *viz.* from the seeds of the last year's fen: Might it not then be adviseable to burn the fenny hop-vines as soon as the hops are picked, in hopes thereby to destroy some of the seed of the mold?

"Mr. *Austin* of *Canterbury* observes fen to be more fatal to those grounds that are low and sheltered, than to the high and open grounds; to those that are shelving to the North, than to the shelving to the South; to the middle of grounds, than to the outsides; to the dry and gentle grounds, than to the moist and stiff grounds. This was very apparent throughout the Plantations, where the land had the same workmanship, and help bestowed upon it, and was wrought at the same time; but if in either of these cases there was a difference, it had a different effect; and the low and gentle grounds, that lay neglected, were then seen less distempered, than the open and moist, that were carefully managed and looked after.

"The honey dews are observed to come about the 11 of *June*, which by the middle of *July* turn the leaves black, and make them stink."

I have in *July* (the season for fire-blasts, as the planters call them) seen the vines in the middle of a hop-ground all scorched up almost from one end of a large ground to the other, when a hot gleam of Sunshine has come immediately after a shower of rain; at which time the vapours are often seen with the naked eye, but especially with reflecting Telescopes, to ascend so plentifully, as to make a clear and distinct object become immediately very dim and tremulous. Nor was there any dry gravelly vein in the ground, along the course of this scorch. It was therefore probably owing to the much greater quantity of scorching vapors, in the middle than outsides of the ground, and that being a denser medium, it was much hotter than a more rare medium.

And perhaps, the great volume of ascending vapor might make the Sun-beams converge a little toward the middle of the ground, that being a denser medium, and thereby increase the heat considerably; for I observed, that the course of the scorched hops was in a line at right angles, to the Sun-beams about a 11 a clock, at which time the hot gleam was: The hop-ground was in a valley which runs from South-west to North-east: And to the best of my remembrance, there was then but little wind, and that in the course of the scorch; but had there been some other gentle wind, either North or South, 'tis not improbable but that the North wind gently blowing the Volume of rising wreak on the

South side of the ground, that side might have been most scorched, and so *vice versâ.*

As to particular fire-blasts, which scorch here and there a few hop-vines, or one or two branches of a tree, without damaging the next adjoining; what *Astronomers* observe, may hint to us as a not very improbable cause of it; *viz.* They frequently observe (especially with the reflecting Telescopes) small separate portions of pellucid vapors floating in the air; which tho' not visible to the naked eye, are yet considerably denser than the circumambient air: And vapors of such a degree of density may very probably, either acquire such a scalding heat from the Sun, as will scorch what plants they touch, especially the more tender: An effect which the gardiners about *London* have too often found to their cost, when they have incautiously put bell-glasses over their Colly-flowers early in a frosty morning, before the dew was evaporated off them; which dew being raised by the Sun's warmth, and confined within the glass, did there form a dense transparent scalding vapor, which burnt and killed the plants. Or perhaps, the upper or lower surface of these transparent separate flying volumes of vapors may, among the many forms they revolve into, sometimes approach so near to a hemisphere, or hemicylinder, as thereby to make the Sun-beams converge enough, often to scorch the more tender plants they shall fall on: And sometimes also, parts of the more hardy plants and trees, in proportion to the greater or less convergency of the Sun's rays.

The learned *Boerhaave*, in his *Theory of Chemistry*, p. 245. observes, "That those white clouds which appear in summer-time, are as it were so many mirrours, and occasion excessive heat. These cloudy mirrours are sometimes round, sometimes concave, polygonous, *&c.* when the face of heaven is covered with such white clouds, the Sun shining among them, must of necessity produce a vehement heat; since many of his rays, which would otherwise, perhaps, never touch our earth, are hereby reflected to us; thus if the Sun be on one side, and the clouds on the opposite one, they will be perfect burning glasses. And hence the phaenomena of thunder.

"I have sometimes (continues he) observed a kind of hollow clouds, full of hail and snow, during the continuance of which the heat was extreme; since by such condensation they were enabled to reflect much more strongly. After this came a sharp cold, and then the clouds discharged their hail in great quantity; to which succeeded a moderate warmth. Frozen concave clouds therefore, by their great reflections, produce a vigorous heat, and the same when resolved excessive cold."

Whence we see that blasts may be occasioned by the reflections of the clouds, as well as by the above mentioned refraction of dense transparent vapors.

July 21. I observed that at that season the top of the Sunflower being tender, and the flower near beginning to blow, that if the Sun rise clear the flower faces towards the East, and the Sun continuing to shine, at noon, it faces to the South; and at 6 in the evening to the West: And this not by turning round with the Sun, but by nutation; the cause of which is, that the side of the stem next the Sun perspiring most, it shrinks, and this plant perspires much.

I have observed the same in the tops of *Jerusalem-artichokes* and of garden-beans in very hot Sun-shine.

Experiment X.

July 27. I fixed an *Apple-branch m*, 3 feet long $\frac{1}{2}$ inch diameter, full of leaves, and lateral shoots to the tube *t*, 7 feet long, $\frac{5}{8}$ diameter. (Fig. 3.) I filled the tube with water, and then immersed the whole branch as far as over the lower end of the tube, into the vessel *uu* full of water.

The water subsided 6 inches the first two hours (being the first filling of the sap vessels) and 6 inches the following night, 4 inches the next day; and $2+\frac{1}{2}$ the following night.

The third day in the morning I took the branch out of the water; and hung it with the Tube affixed to it in the open air; it imbibed this day $27+\frac{1}{2}$ inches in 12 hours.

This Experiment shews the great power of perspiration; since when the branch was immersed in the vessel of water, the 7 feet column of water in the tube, above the surface of the water, could drive very little thro' the leaves, till the branch was exposed to the open air.

This also proves, that the perspiring matter of trees is rather actuated by warmth, and so exhaled, than protruded by the force of the sap upwards.

And this holds true in animals, for the perspiration in them is not always greatest in the greatest force of the blood; but then often least of all, as in fevers.

I have fixed many other branches in the same manner to long tubes, without immersing them in water; which tubes, being filled with water, I could see precisely, by the descent of the water in the tube *t*, how fast it perspired off; and how very little perspired in a rainy day, or when there were no leaves on the branches.

Experiment XI.

Aug. 17. At 11 *a. m*, I cemented to the tube *a b* (Fig. 4.) 9 feet long, and $\frac{1}{2}$ inch diameter an *Apple-branch d* 5 feet long $\frac{6}{8}$ inch diameter; I poured water into the tube, which it imbibed plentifully, at the rate of 3 feet length of the tube in an hour. At 1 a clock I cut off the branch at *c*, 13 inches below the glass-tube. To the bottom of the remaining stem I tyed a glass cistern *z*, covered with ox-gut, to keep any of the water which dropped from the stem *c b* from evaporating. At the same time I set the branch *d r* which I had cut off in a known quantity of water, in the vessel *x*, (Fig. 5.) the branch in the vessel *x* imbibed 18 ounces of water, in 18 hours day and 12 hours night; in which time only 6 ounces of water had passed thro' the stem *c b* (Fig. 4.) which had a column of water 7 feet high, pressing upon it all the time.

This again shews the great power of perspiration; to draw three times more water, in the same time, through the long slender parts of the branch *r* (Fig. 5.) as was pressed thro' a larger stem *c b* (Fig. 4.) of the same branch; but 13 inches long, with 7 feet pressure of water upon it, in the tube *a b*.

I tryed in the same manner another apple-branch, which in 8 hours day imbibed 20 ounces, while only 8 ounces passed thro' the stem *c b*, (Fig. 4.) which had the column of water on it.

The same I tried with a quince branch, which in 4 hours day imbibed 2 ounces $+\frac{1}{3}$, while but $\frac{1}{3}$ ounce passed thro' the stem *c b* (Fig. 4.) which had 9 feet weight of water pressing on it.

Note, All these (under this Experiment 11.) were made the first day, before the stem could be any thing saturate with water, or the sap-vessels shrunk so as to hinder its passage.

Experiment XII.

I cut off from a dwarf *Apple-tree e w* the top of the branch *l*, (Fig. 6.) which was an inch diameter, and fixed to the stem *l*, the glass tube *l b:* then I poured water into the tube, which the branch would imbibe, so as to drink down 2 or 3 pints in a day, especially if I sucked with my mouth at the top of the tube *b*, so as that a few air bubbles were drawn out of the stem *l*; then the water was imbibed so fast, that if I immediately screwed on the mercurial gage, *m r y z*, the mercury would be drawn up to *r*, 12 inches higher than in the other leg.

At another time I poured into the tube *l*, fixed to a golden Renate-tree, a quart of high rectified spirit of wine, camphorated, which quantity the stem imbibed in 3 hours space; this killed one half of the tree: this I did to try if I could give a flavour of camphire to the apples which were in great plenty on the branch. I could not perceive any alteration in the taste of the apples, tho' they hung several weeks after; but the smell of the camphire was very strong in the stalks of the leaves, and in every part of the dead branch.

I made the same experiment on a vine, with strongly-scented orange-flower-water; the event was the same, it did not penetrate into the grapes, but very sensibly into the wood and stalks of the leaves.

I repeated the same experiment on two distant branches of a large Catharine pear-tree, with strong decoctions of sassafras, and of elder-flowers, about 30 days before the pears were ripe; but I could not perceive any taste of the decoctions in the pears.

Tho' in all these cases the sap-vessels of the stem were strongly impregnated with a good quantity of these liquors; yet the capillary sap-vessels near the fruit were so fine, that they changed the texture of, and assimilated to their own substance those high tasted and perfumed liquors; in the same manner as graffs and buds change the very different sap of the stock to that of their own specifick nature.

This experiment may safely be repeated with well scented and perfumed common water, which trees will imbibe at *l l* without any danger of killing them.

Experiment XIII.

In order to try whether the capillary sap-vessels had any power to protrude sap out at their extremities, and in what quantity, I made the three following experiments, *viz.*

In *August* I took a cylinder of an apple-branch, 12 inches long $\frac{7}{8}$ diameter: I set it with its great end downwards in a mint glass, (full of water) tyed over with ox-gut. The top of the stick was moist for 10 days, while another stick of the same branch (but out of water) was very dry. It evaporated an ounce of water in those 10 days.

Experiment XIV.

In *Sept.* I fix'd a tube *t* (Fig. 7.) 7 feet long, to a like stem *s*, as the former, and set the stem in water *x*, to try if, as the water evaporated out of the top of the stem *r*, it would rise to any height in the tube *t*; but it did not rise at all in the tube, tho' the top of the stem was wet: I then filled the tube with water, but it passed freely into the vessel *x*.

Experiment XV.

Sept. 10. $2+\frac{1}{2}$ feet from the ground, I cut off the top of a half standard *Duke Cherry-tree* against a wall, and cemented on it the neck of a Florence flask *f*, (Fig. 8.) and to that flask neck a narrow tube *g*, five feet long, in order to catch any moisture that should arise out of the trunk *y*; but none arose in 4 hours, except a little vapor that was on the flask's neck.

I then dug up the tree by the roots, and set the root in water, with the glasses affixed to the top of the stem; after several hours nothing rose but a little dew, which hung on the inside of *f*; yet it is certain by many of the foregoing experiments, that if the top and leaves of this tree had been on, many ounces of water would in this time have passed thro' the trunk, and been evaporated thro' the leaves.

I have tryed the same experiment with several vine branches cut off, and set in water thus, but no water rose into *f*.

These three last experiments all shew, that tho' the capillary sap vessels imbibe moisture plentifully; yet they have little power to protrude it farther, without the assistance of the perspiring leaves, which do greatly promote its progress.

Experiment XVI.

In order to try whether any sap rose in the winter, I took in *January* several parcels of Filberd-suckers, Vine-branches, green Jessamine-branches, Philarea and Laurel-branches, with their leaves on them, and dipped their transverse cuts in melted cement, to prevent any moisture's evaporating thro' the wounds; I tyed them in separate bundles and weighed them.

The Philberd-suckers decreased in 8 days (some part of which were very wet, but the last 3 or 4 days drying winds) the 11th part of their whole weight.

The vine-cuttings in the same time the $\frac{1}{24}$ part.

The Jessamine in the same time the $\frac{1}{6}$ part.

The Philarea decreased the $\frac{1}{4}$ part in 5 days.

The Laurel the $\frac{1}{4}$ part in 5 days, and more.

Here is a considerable daily waste of sap, which must therefore necessarily be supplied from the root; whence it is plain that some sap rises all the winter, to supply this continual waste, tho' in much less quantity than in summer.

Hence we see good reason why the Ilex and the Cedar of *Libanus* (which were grafted the first on an *English*-oak, the other on the Larix) were verdant all the winter, notwithstanding the Oak and Larix leaves were decayed and fallen off; for tho' when the winter came on, there did not sap enough rise to maintain the Oak and Larix leaves, yet by this present experiment we see, that some sap is continually rising all the winter; and by experiment the 5th on the Limon-tree, and by several other the like experiments, on many sorts of ever-greens, we find that they perspiring little, live and thrive with little nourishment; the Ilex and Cedar might well therefore continue green all the winter, notwithstanding the leaves of the trees they were grafted on fell off. See the curious and industrious Mr. *Fairchild's* account of these graftings in Mr. *Miller's Gardiner's Dictionary. Vol. II. Supplement sap.*

EXPERIMENT XVII.

Having by many evident proofs in the foregoing experiments seen the great quantities of liquor that were imbibed and perspired by trees, I was desirous to try if I could get any of this perspiring matter; and in order to it, I took several glass chymical retorts, *b a p* (Fig. 9.) and put the boughs of several sorts of trees, as they were growing with their leaves on, into the retorts, stoping up the mouth *p* of the retorts with bladder. By this means I got several ounces of the perspiring matter of Vines, Fig-trees, Apple-trees, Cherry-trees, Apricot and Peach-trees; Rue, Horse-radish, Rheubarb, Parsnip, and Cabbage leaves: the liquor of all of them was very clear, nor could I discover any different taste in the several liquors: But if the retort stand exposed to the hot sun, the liquor will taste of the coddled leaves. Its specifick gravity was nearly the same with that of common water; nor did I find many air bubbles in it, when placed in the exhausted receiver, which I expected to have found; but when reserved in open viols, it stinks sooner than common water; an argument that it is not pure water, but has some heterogeneous mixtures with it.

I put also a large Sun-flower full blown, and as it was growing, into the head of a glass-still, and put its rostrum into a bottle, by which means there distilled a good quantity of liquor into the bottle. It will be very easy in the same manner to collect the perspirations of sweet scented Flowers, tho' the liquor will not long retain its grateful odor, but stink in few days.

This experiment would be very proper to begin the learned *Boerhaave's* clear and very rational chymical processes with, as being a degree more simple than his first process, the distillation in a cold still: For this is undisturbed nature's own method of distilling.

EXPERIMENT XVIII.

In order to find out what stores of moisture nature had provided in the earth, (against the dry summer season,) that might answer this great expence of it, which is so necessary for the production and support of vegetables.

July 31. 1724. I dug up a cubick foot earth in an alley which was very little trampled on; weighed (after deducting the weight of the containing vessel) 104 pounds + 4 ounces + $\frac{1}{3}$. A cubick foot of water weighs 59 + $\frac{1}{2}$ which is little more than half the specifick gravity of earth. This was a dry season, with a mixture of some few showers, so that the grass-plat adjoyning was not burnt up.

At the same time I dug up another cubick foot of earth, from the bottom of the former, it weighed 106 pound + 6 ounces, + $\frac{1}{3}$.

I dug up also a third cubick foot of earth, at the bottom of the two former, it weighed 111 pounds + $\frac{1}{3}$.

These three feet depth were a good brick earth, next to which was gravel, in which at 2 feet depth, *viz.* 5 feet below the surface of the earth, the springs did then run.

When the first cubick foot of earth was so dry and dusty, as to be unfit for vegetation I weighed it, and found it had lost 6 pound + 11 ounces, or 194 cubick inches of water, near $\frac{1}{8}$ part of its bulk.

Some days after, the second cubick foot being dryer than either the first or third, was decreased in weight 10 pounds.

The third cubick foot, being very dry and dusty, had lost 8 pounds + 8 ounces, or 247 cubick inches, *viz.* $\frac{1}{7}$ part of its bulk.

Now supposing the roots of the Sunflower (the longest of which reached 15 inches every way from the stem) to occupy and draw nourishment from 4 cubick feet of earth, and suppose each cubick foot of earth to afford 7 pounds of moisture, before it be too dry for vegetation; the Plant imbibing and perspiring 22 ounces every 24 hours, that will be 28 pounds of water, which will be drawn off in 21 days and 6 hours; after which the Plant would perish if there were not fresh supplies to these 4 cubick feet of earth, either from dew or moisture arising from below 15 inches (the depth of the roots) up into the earth occupied by the roots.

Experiment XIX.

In order to find out the quantity of *Dew* that fell in the night, *Aug.* 15. at 7. *p.m.* I chose two glazed earthen Pans which were 3 inches deep, and 12 inches diameter in surface; I filled them with pretty moist earth taken off the surface of the earth; they increased in weight by the night's dew 180 grains, and decreased in weight by the evaporation of the day 1 ounce + 282 grains.

N. B. I set these Pans in other broader Pans, to prevent any moisture from the earth sticking to the bottoms of them. The moister the earth, the more Dew there falls on it in a night, and more than a double quantity of Dew falls on a surface of water, than there does on an equal surface of moist earth. The evaporation of a surface of water in 9 hours winter's dry day is $\frac{1}{21}$ of an inch. The evaporation of a surface of Ice, set in the shade during nine hours day, was $\frac{1}{31}$.

So here are 540 grains more evaporated from the earth every 24 hours in summer, than falls in Dew in the night; that is, in 21 days near 26 ounces, from a circular area of a foot diameter; and circles being as the squares of their diameters 10 pounds + 2 ounces, will in 21 days be evaporated from the hemisphere of 30 inches diameter, which the Sunflower's root occupies: Which with the 29 pounds drawn off by the Plant in the same time, makes 39 pounds, that is, 9 pounds and $\frac{3}{4}$ out of every cubick foot of earth, the Plant's roots occupying more than 4 cubick feet; but this is a much greater degree of dryness than the surface of the earth ever suffers for 15 inches depth, even in the dryest seasons in this country.

In a long dry season, therefore, especially within the Tropicks, we must have recourse for sufficient moisture (to keep Plants and Trees alive) to the moist strata of earth, which lay next below that in which the roots are. Now moist bodies always communicate of their moisture to more dry adjoyning bodies; but this slow motion of the ascent of moisture is much accelerated by the Sun's heat to considerable depths in the earth, as is probable from the following 20th experiment.

Now 180 grains of Dew falling in one night, on a circle of a foot diameter, $=113$ square inches; these 180 grains being equally spread on this surface, its depth will be $\frac{1}{159}$ part of an inch $=\frac{180}{113\times 254}$. I found the depth of Dew in a winter night to be the $\frac{1}{90}$ part of an inch; so that if we allow 151 nights for the extent of the summer's Dew, it will in that time arise to one inch depth. And reckoning the remaining 214 nights for the extent of the winter's Dew, it will produce 2. 39 inches depth, which makes the Dew of the whole year amount to 3. 39 inches depth.

And the quantity which evaporated in a fair summer's day from the same surface, being 1 ounce $+$ 282 grains, gives $\frac{1}{40}$ part of an inch depth for evaporation, which is four times as much as fell at night.

I found, by the same means, the evaporation of a winter's day to be nearly the same as in a summer's day; for the earth being in winter more saturate with moisture, that excess of moisture answers to the excess of heat in summer.

Nic. Cruquius No 381. of the Philosophical Transactions, found that 28 inches depth evaporated in a whole year from water, *i.e.* $\frac{1}{12}$ of an inch each day, at a mean rate; but the earth in a summer's day evaporates $\frac{1}{40}$ of an inch; so the evaporation of a surface of water, is to the evaporation of a surface of earth in summer, as 10 : 3.

The quantity of Rain and Dew which falls in a year is at a medium 22 inches: The quantity of the earth's evaporation in a year is at least $9+\frac{1}{2}$ inches, since that is the rate, at which it evaporates in a summer's day: From which $9+\frac{1}{2}$ inches is to be deducted 3. 39 inches for circulating daily Dew; there remains 6. 2 inches, which 6. 2 inches deducted from the quantity of Rain which falls in a year, there remains at least 16 inches depth, to replenish the earth with moisture for vegetation, and to supply the Springs and Rivers.

In the case of the hop-ground, the evaporation from the hops may be considered only for 3 months at $\frac{1}{101}$ part of an inch each day, which will be $\frac{9}{10}$ of an inch; but before we allow 6. 2 inches vapor to evaporate from the surface of the ground, which added to $\frac{9}{10}$ inch gives 7. 1 inches, which is the utmost that can be evaporated from a surface of hop-ground in a year. So that of 22 inches depth of rain, there remains 15 inches to supply springs; which are more or less exhausted, according to the dryness or wetness of the year. Hence we find that 22 inches depth of rain in a year is sufficient for all the purposes of nature, in such flat countries as this about *Teddington* near *Hampton Court*. But in the hill countries, as in *Lancashire*, there falls 42 inches depth of rain-water; from which deducting 7 inches for evaporation, there remains 35 inches depth of water for the springs; besides great supplies from much more plentiful dews, than fall in plain countries: Which vast stores seem so abundantly sufficient to answer the great quantity of water, which is conveyed away, by springs and rivers, from those hills, that we need not have recourse, for supplies, to the great *Abyss*, whose surface, at high water, is surmounted some hundreds of feet by ordinary hills, and some thousands of feet by those vast hills from whence the longest and greatest rivers take their rise.

Experiment XX.

I provided me six *Thermometers*, whose stems were of different lengths, *viz.* from 18 inches to 4 feet. I graduated them all by one proportional scale, beginning from the freezing point; which may well be fixed as the utmost boundary of vegetation on the side of cold, where the work

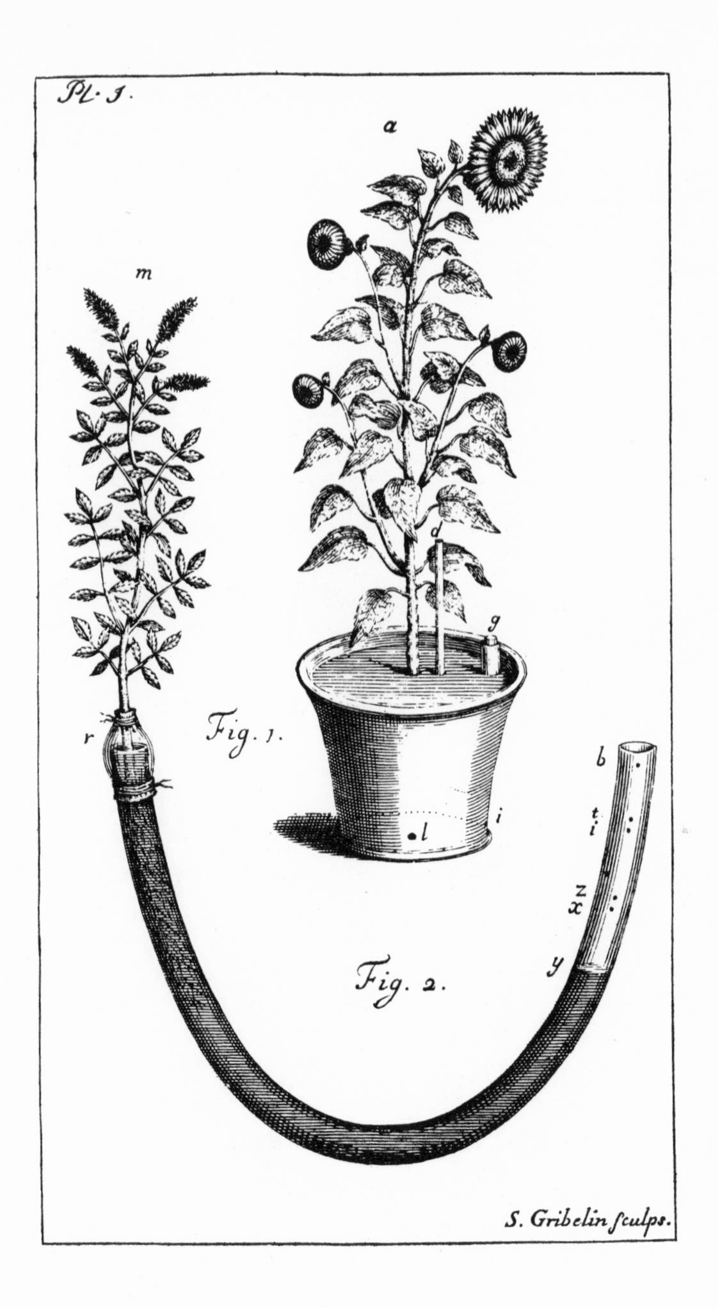
Pl. I.
a
m
d
g
r
Fig. 1.
b
t
i
i
l
z
x
y
Fig. 2.
S. Gribelin sculps.

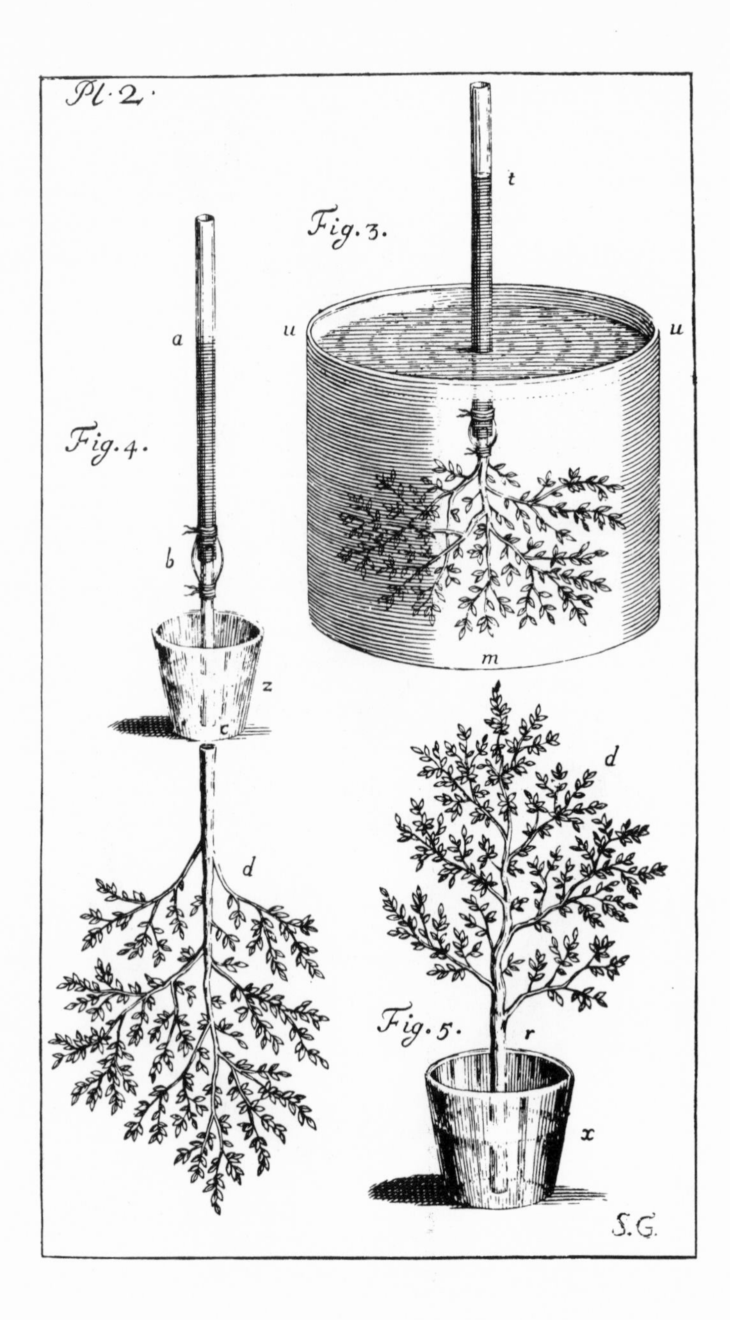
Pl. 2.
Fig. 3.
t
u
u
m
Fig. 4.
a
b
z
c
d
Fig. 5.
d
r
x
S.G.

of vegetation ceases, the watry vehicle beginning then to condense and be fixed; tho' many trees, and some plants, as grass, moss, *&c.* do survive it; yet they do not vegetate at that time.

The greatest degree of heat, which I marked on my *Thermometers*, was equal to that of water, when heated to the greatest degree that I could bear my hand in it, without stirring it about. A degree of heat, which is the middle, between the freezing point, and the heat of boiling water, which being too great for vegetation, may therefore be fixed, as the utmost boundary of vegetation, on the warm side; beyond which plants will rather fade than vegetate, such a degree of heat separating and dispersing, instead of congregating, and uniting the nutritive particles.

This space I divided into 90 degrees on all the *Thermometers*, beginning to number from the freezing point. Sixty four of these degrees is nearly equal to the heat of the blood of animals; which I found by the rule given in the *Philosophical Transactions*, Vol. II. p. 1. of Mr. *Motte's Abridgment,* viz. by placing one of the *Thermometers* in water heated to the greatest degree, that I could bear my hand in it, stirring it about: And which I was further assured of, by placing the ball of my *Thermometer* in the flowing blood of an expiring Ox. The heat of the blood to that of boiling water is as $14+\frac{3}{11}$ to

By placing the ball of one of these *Thermometers* in my bosom, and under an armpit, I found the external heat of the body 54 of these degrees. The heat of milk, as it comes from the Cow, is 55 degrees, which is nearly the same with that for hatching of eggs. The heat of urine 58 degrees. The common temperate point in *Thermometers* is about 18 degrees.

The hottest Sun-shine in the year 1724, gave to the *Thermometer*, exposed to it, a heat equal to that of the blood of animals, *viz.* 64 degrees: And tho' plants endure this and a considerably greater heat, within the tropicks, for some hours each day, yet the then hanging of the leaves of many of them shews that they could not long subsist under it, were they not frequently refreshed by the succeeding evening and night.

The common noon-tide heat in the Sun in *July* is about 50 degrees: The heat of the air in the shade in *July* is at a medium 38 degrees. The *May* and *June* heat is from 17 to 30 degrees; the most genial heat for the generality of plants, in which they flourish most, and make the greatest progress in their growth. The autumnal and vernal heat may be reckoned from 10 to 20 degrees. The winter heat from the freezing point to 10 degrees.

The scorching heat of a hot-bed of horse-dung, when too hot for plants, is 75 degrees and more, and hereabout is probably the heat of blood in high fevers.

The due healthy heat of a hot-bed of horse-dung, in the fine mold, where the roots of thriving Cucumber-plants were, in *Feb.* was 56 degrees, which is nearly the bosom heat, and that for hatching of eggs. The heat of the air under the glass-frame of this hot-bed was 34 degrees; so the roots had 26 degrees more heat, than the plants above ground. The heat of the open air was then 17 degrees.

It is now grown a common and very reasonable practice, to regulate the heat of stoves and green-houses, by means of *Thermometers*, hung up in them. And for greater accuracy, many have the names of some of the principal exoticks written upon their *Thermometers*, over-against, the several degrees of heat, which are found by experience to be properest for them. And I am informed that many of the most curious Gardiners about *London* have agreed to make use of *Thermometers* of this sort; which are made by Mr. *John Fowler* in *Swithins-alley*, near the *Royal-Exchange*; which have the names of the following plants, opposite to their respective most kindly degrees of heat; which in my *Thermometers* answer nearly to the following degrees of heat above the freezing point, *viz.* Melon-thistle 31, Ananas

29, Piamento 26, Euphorbium 24, Cereus 21½, Aloe 19, Indian-fig 16½ Ficoides 14, Oranges 12, Myrtles 9.

Mr. *Boyle*, by placing a *Thermometer* in a cave which was cut strait into the bottom of a cliff, fronting the Sea, to the depth of 130 feet, found the spirit stood, both in winter and summer, at a small division above temperate; the cave had 80 feet depth of earth above it. *Boyle*'s Works, Vol. III. p. 54.

I marked my 6 *Thermometers* numerically, 1, 2, 3, 4, 5, 6. The Thermometer numb. 1. which was shortest, I placed with a South aspect, in the open air; the ball of numb. 2, I set two inches under ground; that of numb. 3, four inches under ground; numb. 4, 8 inches; numb. 5, 16 inches; and numb. 6, 24 inches under ground. And that the heat of the earth, at these several depths, may the more accurately be known, it is proper to place near each *Thermometer* a glass-tube sealed at both ends, of the same length with stems of the several *Thermometers*; and with tinged spirit of wine in them, to the same height, as in each corresponding *Thermometer*; the scale of degrees, of each *Thermometer*, being marked on a sliding ruler, with an index at the back of it, pointing to the corresponding tube. When at any time an observation is to be made, by moving the index, to point to the top of the spirit in that tube, an accurate allowance is hereby made, for the very different degrees of heat and cold, on the stems of the *Thermometers*, at all depths; by which means the scale of degrees will shew truly the degrees of heat in the balls of the *Thermometers*, and consequently, the respective heats of the earth, at the several depths where they are placed. The stems of these *Thermometers*, which were above ground, were fenced from weather and injuries, by square wooden tubes; the ground they were placed in was a brick earth in the middle of my garden.

July 30. I began to keep a register of their rise and fall. During the following month of *August*, I observed that when the spirit in the Thermometer numb. 1, (which was exposed to the Sun) was about noon risen to 48 degrees, then the second Thermometer was 45 degrees, the 5th 33, and the 6th 31; the 3rd and 4th at intermediate degrees. The 5th and 6th Thermometers kept nearly the same degree of heat, both night and day, till towards the latter end of the month; when as the days grew shorter and cooler, and the nights longer and cooler, they then fell to 25 and 27 degrees.

Now, so considerable a heat of the Sun, at two feet depth, under the earth's surface, must needs have a strong influence in raising the moisture at that and greater depths; whereby a very great and continual wreak must always be ascending, during the warm summer season, by night as well as day; for the heat at two feet depth is nearly the same night and day: The impulse of the Sun-beams giving the moisture of the earth a brisk undulating motion, which watery particles, when separated and rarified by heat, do ascend in the form of vapour: And the vigour of warm and confined vapour, (such as is that which is 1, 2, or 3 feet deep in the earth) must be very considerable, so as to penetrate the roots with some vigour; as we may reasonably suppose, from the vast force of confined vapor in *Æolipiles*, in the digester of bones, and the engine to raise water by fire.

If plants were not in this manner supplied with moisture, it were impossible for them to subsist, under the scorching heats, within the tropicks, where they have no rain for many months together: For tho' the dews are much greater there, than in these more Northern climates; yet doubtless where the heat so much exceeds ours, the whole quantity evaporated in a day there, does as far exceed the quantity that falls by night in dew, as the quantity evaporated here in a summer's day, is found to exceed the quantity of dew which falls in the night. But the dew, which falls in a hot summer season, cannot possibly be of any benefit to the roots of trees; because it is remanded back from the earth, by the following day's heat, before so small a quantity of moisture can have soaked to any considerable depth. The great benefit therefore of dew, in hot weather, must be, by being plentifully imbibed into vegetables; thereby not only refreshing them for the present, but also furnishing them with a fresh supply of moisture towards the great expences of the succeeding day.

'Tis therefore probable, that the roots of trees and plants are thus, by means of the Sun's warmth, constantly irrigated with fresh supplies of moisture; which, by the same means, insinuates it self with some vigour into the roots. For if the moisture of the earth were not thus actuated, the roots must then receive all their nourishment meerly by imbibing the next adjoining moisture from the earth; and consequently the shell of earth, next the surface of the roots, would always be considerably drier the nearer it is to the root; which I have not observed to be so. And by *Exper.* 18, and 19, the roots would be very hard put to it, to imbibe sufficient moisture in dry summer weather, if it were not thus conveyed to them, by the penetrating warmth of the Sun: Whence by the same genial heat, in conjunction with the attraction of the capillary sap vessels, it is carried up thro' the bodies and branches of vegetables, and thence passing into the leaves, it is there most vigorously acted upon, in those thin plates, and put into an undulating motion, by the Sun's warmth, whereby it is most plentifully thrown off, and perspired thro' their surface; whence, as soon as it is disintangled, it mounts with great rapidity in the free air.

But when, towards the latter end of *October*, the vigour of the Sun's influence is so much abated, that the first *Thermometer* was fallen to 3 degrees above the freezing point, the second to 10 degrees, the fifth to 14 degrees, and the sixth *Thermometer* to 16 degrees; then the brisk undulations of the moisture of the earth, and also of the ascending sap, much abating, the leaves faded and fell off.

The greatest degree of cold, in the following winter, was in the first 12 days of *November*; during which time, the spirit in the first *Thermometer* was fallen 4 degrees below the freezing point, the deepest *Thermometer* 10 degrees, the ice on ponds was an inch thick. The Sun's greatest warmth, at the winter solstice, in a very serene, calm, frosty-day, was, against a South aspect of a wall, 19 degrees, and in a free open air, but a 11 degrees above the freezing point. From the 10th of *January* to the 29th of *March* was a very dry season; when the green Wheat was generally the finest that was ever remembred. But from the 29th of *March* 1725, to the 29th of *September* following, it rained more or less almost every day, except 10 or 12 days about the beginning of *July*; and that whole season continued so very cool, that the spirit in the first *Thermometer* rose but to 24 degrees, except now and then in a short interval of Sunshine; the second only to 20 degrees; the fifth and sixth to 24 and 23 degrees, with very little variation: So that during this whole summer, those parts of roots which were two feet under ground, had 3 or 4 degrees more warmth than those which were but two inches under ground: And at a medium the general degree of heat thro' this whole summer, both above and under ground, was not greater than the heat of the middle of the preceding *September*.

The year 1725, having been both in this Island, and in the neighbouring Nations, most remarkably wet and cold; and the year 1723, in the other extream, as remarkably dry, as has ever been known; it may not be improper here to give a short account of them, and the influence they had on their productions.

"Mr. *Miller*, in the account which he took of the year 1723, observed that the winter was mild and dry, except that in *February* it rained almost every day, which kept the spring backward. *March, April, May, June*, to the middle of *July*, proved extreamly dry, the wind North-east most part of the time. The fruits were forward, and pretty good; but kitchen-stuff, especially Beans and Pease, failed much. The latter half of *July* the weather proved very wet, which caused the fruits to grow so fast, that many of them rotted on the trees; so that the autumn fruits were not good. There were great plenty of Melons, very large, but not well tasted. Great plenty of Apples; many kinds of fruits blossomed in *August*, which produced many small Apples and Pears in *October*, as also Strawberries and Raspberries in great plenty. Wheat was good, little Barley, much of which was very unequally ripe, some not at all, because sown late, and no timely rain to fetch it up. There were innumerable Wasps; how it fared with the hops this dry year, is mentioned under *Exper. 9*.

"The following winter 1724, proved very mild; the spring was forward in *January*, so that the *Snow-drops, Crocus's, Polyanthus's, Hepatica's,* and *Narcissus's,* were in Flower. And it was remarkable, that most of the Colliflower plants were destroyed by the mildew, of which there was more, all this winter, than had been known in the memory of man. In *February* we had cold sharp weather, which did some damage to the early crops, and it continued variable till *April*; so that much of the early Wall-fruit was cut off: And again the 6th of *May* was a very sharp frost, which much injured tender plants and fruits. The summer in general was moderately dry, the common fruits proved pretty good, but late: Melons and Cucumbers were good for little: Kitchen-stuff was in great plenty in the markets."

In the very wet and cold year 1725, most things were a full month backwarder than usual. Not half the Wheat in by the 24th of *August*, in the *Southern* parts of *England*; very few Melons or Cucumbers, and those not good. The tender Exoticks fared but ill; scarce any Grapes, those small, and of very unequal sizes, on the same bunch, not ripe; Apples and Pears green and insipid; no fruit nor products of the ground good, but crude: Pretty good plenty of Wheat tho' coarse, and long straw; Barley coarse, but plenty of it in the uplands. Beans and Pease, most flourishing and plentiful; few Wasps or other insects, except Flies on hops. Hops were very bad thro' the whole Kingdom. Mr. *Austin* of *Canterbury* sent me the following particular account, how it far'd with them there; where they had more than at *Farnham*, and most other places, *viz.*

"At mid-*April* not half the shoots appeared above ground; so that the planters knew not how to pole them to the best advantage. This defect of the shoot, upon opening the hills, was found to be owing to the multitude and variety of vermin that lay preying upon the root; the increase of which was imputed to the long and almost uninterrupted series of dry weather, for three months past: Towards the end of *April*, many of the hop-vines were infested with the Flies. About the 20th of *May* there was a very unequal crop, some Vines being run seven feet, others not above three or four feet; some just tied to the poles, and some not visible: And this disproportionate inequality in their size continued thro' the whole time of their growth. The Flies now appeared upon the leaves of the forwardest Vines, but not in such numbers here, as they did in most other places. About the middle of *June*, the Flies increased, yet not so as to endanger the crop; but in distant plantations they were exceedingly multiplied, so as to swarm towards the end of the month. *June* 27th some specks of fen appeared: From this day to the 9th of *July*, was very fine dry weather. At this time, when it was said that the hops in most other parts of the Kingdom looked black and sickly, and seemed past recovery, ours held it out pretty well, in the opinion of the most skilful Planters. The great leaves were indeed discoloured, and a little withered, and the fen was somewhat increased. From the 9th of *July* to the 23rd the Fen increased a good deal, but the Flies and Lice decreased, it raining daily much: In a week more the Fen, which seemed to be almost at a stand, was considerably increased, especially in those grounds where it first appeared. About the middle of *August*, the Vines had done growing both in stem and branch; and the forwardest began to be in Hop, the rest in Bloom: the Fen continued spreading, where it was not before perceived, and not only the leaves, but many of the Burrs also were tainted with it. About the 20th of *August*, some of the Hops were infected with the Fen, and whole branches corrupted by it. Half the Plantations had hitherto pretty well escaped, and from this time the Fen increased but little: But several days violent wind and rain, in the following week, so disordered them, that many of them began to dwindle, and at last came to nothing; and of those that then remained in bloom, some never turned to Hops; and of the rest which did, many of them were so small, that they very little exceeded the bigness of a good thriving Burr. We did not begin to pick till the 8th of *September*, which was 18 days later than we began the year before: The crop was little above two hundred on an acre round, and not good." The best Hops sold this year at *Way-Hill* Fair for sixteen pounds the hundred.

The almost uninterrupted wetness and coldness of the year 1725, very much affected the produce of the Vines the ensuing year; and we have sufficient proof from the observations that the 4 or 5 last years afford us, that the moisture or dryness of the preceding year, has a considerable influence on the productions of the Vine the following year. Thus in the year 1722, there was a dry season, from the beginning of *August* thro' the following autumn and winter, and the next summer there was good plenty of Grapes. The year 1723 was a remarkably dry year, and in the following year 1724, there was an unusual plenty of Grapes. The year 1724 was moderately dry, and the following spring the Vines produced a sufficient quantity of bunches, but by reason of the wetness and coldness of the year 1725 they proved abortive, and produced hardly any Grapes. This very wet year had an ill effect, not only upon its own productions, but also on those of the following year: For notwithstanding there was a kindly spring and blooming season in the year 1726, yet there were few bunches produced, except here and there in some very dry soils. This many Gardiners foresaw early, when upon pruning of the Vines, they observed the bearing shoots to be crude and immature; which was the reason why they were not fruitful. The first crop thus failing in many places, the Vines produced a second, which had not time to come to maturity before the cold weather came on.

I have often observed from these Thermometers, when that kind of hovering lambent Fog arises, (either mornings or evenings) which frequently betokens fair weather, that the air which in the preceding day was much warmer, has upon the absence of the sun become many degrees cooler than the surface of the earth; which being near 1500 times denser than the air, cannot be so soon affected with the alternacies of hot and cold; whence 'tis probable, that those vapours which are raised by the warmth of the earth, are by the cooler air soon condensed into a visible form. And I have observed the same difference between the coolness of the air, and the warmth of water in a pond, by putting my Thermometer, which hung all night in the open air in summer time, into the water, just before the rising of the sun, when the like reek or fog was rising on the surface of the water.

CHAP. II.

Experiments, whereby to find out the force with which Trees imbibe moisture.

Having in the 1st chapter seen many proofs of the great quantities of liquor imbibed and perspired by vegetables, I propose in this, to enquire with what force they do imbibe moisture.

Tho' vegetables (which are inanimate) have not an engine, which, by its alternate dilatations and contractions, does in animals forcibly drive the blood through the arteries and veins; yet has nature wonderfully contrived other means, most powerfully to raise and keep in motion the sap, as will in some measure appear by the experiments in this and the following chapter.

I shall begin with an experiment upon roots, which nature has providently taken care to cover with a very fine thick strainer; that nothing shall be admitted into them, but what can readily be carried off by perspiration, vegetables having no other provision for discharging their recrement.

EXPERIMENT XXI.

August 13. in the very dry year 1723, I dug down $2+\frac{1}{2}$ feet deep to the root of a thriving baking *Pear-tree*, and laid bare a root $\frac{1}{2}$ inch diameter *n*. (Fig. 10.) I cut off the end of the root at *i*, and put the remaining stump *i n* into the glass tube *d r*, which was an inch diameter, and 8 inches long, cementing it fast at *r*; the lower part of the tube *d z* was 18 inches long, and $\frac{1}{4}$ inch diameter in bore.

Then I turned the lower end of the tube *z* uppermost, and filled it full of water, and then immediately immersed the small end *z* into the cistern of mercury *x*; taking away my finger, which stopped up the end of the tube *z*.

The root imbibed the water with so much vigor, that in 6 minutes time the mercury was raised up the tube *d z* as high as *z*, *viz*. 8 inches.

The next morning at 8 a clock, the mercury was fallen to 2 inches height, and 2 inches of the end of the root *i* were yet immersed in water. As the root imbibed the water, innumerable air bubbles issued out at *i*, which occupied the upper part of the tube at *r* as the water left it.

Experiment XXII.

The eleventh experiment shews, with what great force branches imbibe water, where a branch with leaves imbibed much more than a column of 7 feet height of water could in the same time drive thro' 13 inches length of the biggest part of its stem. And in the following experiments we shall find a further proof of their strong imbibing power.

May 25, I cut off a branch of a young thriving *Apple-tree* *b*, (Fig. 11.) about 3 feet long, with lateral branches; the diameter of the transverse cut *i*, where it was cut off, was $\frac{3}{4}$ of an inch: The great end of this branch I put into the cylindrical glass *e r*, which was an inch diameter within, and eight inches long.

I then cemented fast the joynt *r*, first folding a strap of sheeps skin round the stem, so as to make it fit well to the tube at *r*; then I cemented fast the joynt with a mixture of Bees-wax and turpentine melted together in such a proportion, as to make a very stiff clammy Paste when cold, and over the cement I folded several times wet Bladders, binding it firm with Pack-thread.

At the lower end of the large tube *e* was cemented, on a lesser tube *z e*, $\frac{1}{4}$ inch diameter in bore, and 18 inches long: The substance of this tube ought to be full $\frac{1}{8}$ of an inch thick, else it will too easily break in making this experiment.

These two tubes were cemented together at *e*, first with common hard brick-dust cement to keep the tubes firm to each other; but this hard cement would, by the different dilatations and contractions of the glass and cement, separate from the glass in hot weather, so as to let in air; to prevent which inconvenience, I further secured the joynt with the cement of Bees-wax and Turpentine, binding a wet bladder over all.

When the branch was thus fixed, I turned it downwards, and the glass tube upwards, and then filled both tubes full of water; upon which I immediately applied the end of my finger to close up the end of the small tube, and immersed it as fast as I could into the glass cistern *x*, which was full of mercury and water.

When the branch was now uppermost, and placed as in this figure, then the lower end of the branch was immersed 6 inches in water, *viz.* from *r* to *i*.

Which water was imbibed by the branch, at its transverse cut *i*; and as the water ascended up the sap vessels of the branch, so the mercury ascended up the tube *e z* from the cistern *x*; so as in half an hour's time the mercury was risen 5 inches and $\frac{3}{4}$ high up to *z*.

And this height of the mercury did in some measure shew the force with which the sap was imbibed, tho' not near the whole force; for while the water was imbibing, the transverse cut of the branch was covered with innumerable little hemispheres of air, and many air bubbles issued out of the sap vessels, which air did in part fill the tube *e r*, as the water was drawn out of it; so that the height of the mercury could only be proportionable to the excess of the quantity of water drawn off, above the quantity of air which issued out of the wood.

And if the quantity of air, which issued from the wood into the tube, had been equal to the quantity of water imbibed, then the mercury would not rise at all; because there would be no room for it in the tube.

But if 9 parts in 12 of the water be imbibed by the branch, and in the mean time but 3 such parts of air issue into the tube, then the mercury must needs rise near 6 inches, and so proportionably in different cases.

I observed in this, and most of the following experiments of this sort, that the mercury rose highest, when the sun was very clear and warm; and towards evening it would subside 3 or 4 inches, and rise again the next day as it grew warm, but seldom to the same height it did at first. For I have always found the sap vessels grow every day, after cutting, less pervious, not only for water, but also for the sap of the vine, which never passes to and fro so freely thro' the transverse cut, after it has been cut 3 or 4 days, as at first; probably, because the cut capillary vessels are shrunk, the vesicles also, and interstices between them, being saturate and dilated with extravasated sap, much more than they are in a natural state.

If I cut an inch or two off the lower part of the stem, which has been much saturated by standing in water, then the branch will imbibe water again afresh; tho' not altogether so freely, as when the branch was first cut off the tree.

I repeated the same experiment as this 22d, upon a great variety of branches of several sizes and of different kinds of trees, some of the principal of which are as follow, *viz.*

Experiment XXIII.

July 6th and 8th, I repeated the same experiment with several green shoots *of the Vine*, of this year's growth, each of them full two yards long.

The mercury rose much more leisurely in these experiments, than with the Apple-tree branch; the more the sun was upon it, the faster and higher the mercury rose, but the Vine-branches could not draw it above 4 inches the first day, and 2 inches the third day.

And as the sun set, the mercury sometimes subsided wholly, and would rise again the next day, as the sun came on the Vine-branch.

And I observed, that where some of these Vine-branches were fix'd on the north-side of the large trunk of a Pear-tree, the mercury then rose most in the evening about 6 a clock, as the sun came on the Vine-branch.

Experiment XXIV.

August 9, at 10 *ante Merid.* (very hot sunshine) I fixed in the same manner as *Ex.* 22. a Non-pareil branch, which had 20 Apples on it; it was 2 feet high, with lateral branches, its transverse cut $\frac{5}{8}$ inch diameter: It immediately began to raise the mercury most vigorously, so as in 7 minutes it was got up to *z* 12 inches high.

Mercury being $13+\frac{2}{3}$ times specifically heavier than water, it may easily be estimated to what height the several branches in these experiments would raise water; for if any branch can raise mercury 12 inches, it will raise water 13 feet+ 8 inches: A further allowance being also made for the perpendicular height of the water in the tubes, between *r* and *z* the top of the column of mercury, for that column of water is lifted up by the mercury, be it more or less.

At the same time, I tryed a Golden Renate branch 6 feet long, the mercury rose but 4 inches, it rising higher or lower in branches nearly of the same size and of the same kind of tree, according as the air issued thro' the stem, more or less freely. In the preceding experiment on the Non-pareil branch, I had sucked a little with my mouth at the small end of the tube, to get some air bubbles out of it, before I immersed it in the mercury; (but these air bubbles are best got out by a small wire run to and fro in the tube) and this suction made air bubbles arise out of the transverse cut of the branch; but tho' the quantity of those air bubbles thus sucked out, was but small; yet in this and many other experiments, I found that after such suction, the water was imbibed by the branch, much more greedily, and in much greater quantity than the bulk of the air was, which was sucked out. Probably therefore, these air bubbles, when in the sap vessels, do stop the free ascent of the water, as is the case of little portions of air got between the water in capillary glass tubes.

When the mercury is raised to its greatest height, by precedent suction with the mouth, (which height it reaches sometimes in 7 minutes, sometimes in half an hour or an hour) then from that time it begins to fall, and continues so to do, till it is fallen 5 or 6 inches, the height the branch would have drawn it to, without sucking with the mouth.

But when in a very warm day, the mercury is drawn up 5 or 6 inches, (without precedent suction with the mouth) then it will usually hold up to that height for several hours, *viz.* during the vigorous warmth of the sun; because the sun is all that time strongly exhaling moisture from the branch thro' the leaves, on which account it must therefore imbibe water the more greedily, as is evident by many experiments in the first chapter.

When a branch is fixed to a glass tube set in mercury, and the mercury subsides at night, it will not rise the next morning (as the warmth of the sun increases upon it) unless you fill the tube first full of water: For if half or $\frac{1}{4}$ of the large tube *c r* be full of air, that air will be rarified by the sun; which rarefaction will depress the water in the tube, and consequently the mercury cannot rise.

But where little water is imbibed the first day, (as in the case of the green shoots of the Vine, Exper. XXIII.) then the mercury will rise the second and third day, as the warmth of the sun comes on, without refilling the little water that was imbibed.

Experiment XXV.

In order to make the like experiment on larger branches (when I expected the mercury would have risen much higher than in small ones) I caused glasses to be blown of the shape of this here described (Fig. 12.) of several dimensions at *r*, from two to five inches diameter, with a proportionably large cavity *c:* the stem *z* as near $\frac{1}{4}$ inch diameter as could be, the length of the stem 16 inches.

I cemented one of these glass vessels to a large smooth barked thriving branch of an *Apple-tree*, which was 12 feet long, $1 + \frac{3}{4}$ inch diameter at *i:* I filled the glass tube with water, and immersed the small end in the mercury *x*, which rose but 4 inches, yet it imbibed water plentifully; but the air issued too fast out of the branch at *i*, for the mercury to rise high.

This, and many other experiments of this kind, convince me that branches of 2, 3, or 4 years old, are the best adapted to draw the mercury highest: The vessels of those that are older being too large and pervious to the air, which passes most freely thro' the bark, especially at old eyes: As will be more fully proved in the fifth chapter.

Experiment XXVI.

July 30th at noon, a mixture of sun and clouds, the day and night before, 24 hours continual rain: I cut off a branch of a *Golden Pippin-tree b b* (Fig. 13.) about 3 feet long, with several large lateral branches; its diameter at the great end *p* near an inch, which end I cemented well, and tyed over it a piece of wet bladder.

Then I cut off at *i* the main top twig, where it was $\frac{1}{2}$ inch diameter: I cemented the glass tube *z r*, to the remaining branch *i r*, and then filling the tube with water, set its lower end in the mercury *x*: So that now the branch was placed with its top *i* downwards in the water, in the Aqueo-mercurial gage.

It imbibed the water with such strength, as to raise the mercury with an almost equable progression $11 + \frac{1}{2}$ inches by 3 a clock, (the sun shining then very warm) at which time the water in the tube *r i* being all imbibed; so that the end *i* of the branch was out of the water, then the air bubbles passing more freely down to *i*, and no water being imbibed, the mercury subsided 2 or 3 inches in an hour.

At a quarter past 4 a clock, I refilled the gage with water, upon which the mercury rose afresh from the cistern, *viz.* 6 inches the first $\frac{1}{4}$ of an hour, and in an hour more the mercury reached the same height as before, *viz.* $11 + \frac{1}{2}$ inches. And in an hour and $\frac{1}{4}$ more it rose $\frac{1}{4}$ inch more than at first; but in half an hour after this it began gently to subside; *viz.* because the sun declining and setting, the perspiration of the leaves decreased, and consequently the imbibing of the water at *i* abated, for the end *i* was then an inch in water.

July 31st. It raining all this day, the mercury rose but 3 inches, which height it stood at all the next night. *August* 1st fair sun-shine; this day the mercury rose to 8 inches: This shews again the influence of the sun, in raising the mercury.

This Experiment proves that branches will strongly imbibe from the small end immersed in water to the great end; as well as from the great end immersed in water to the small end; and of this we shall have further proof in the fourth chapter.

Experiment XXVII.

In order to try, whether branches would imbibe with the like force, with the bark off, I took two branches which I call *M* and *N*; I fixed *M* in the same manner as the branch in the foregoing Experiment, with its top downwards, but first I took off all the bark from *i* to *r*. Then I fix'd in the same manner the branch *N*, but with its great end downwards, having also taken off all the bark from *i* to *r*; both the branches drew the mercury up to *z*, 8 inches; so they imbibed with equal strength at either end, and that without bark.

Experiment XXVIII.

August 13. I stripped the leaves off an *Apple tree branch*, and then fixed the great end of the stem in the gage; it raised the mercury $2+\frac{1}{2}$ inches, but it soon subsided, for want of the plentiful perspiration of the leaves, so that the air came in almost as fast as the branch imbibed water.

Experiment XXIX.

I tryed also with what force branches would imbibe, at their small ends, as they are in their natural state growing to the trees.

August 2d I cemented fast the gage *r i z* (Fig. 14) to the pliant branch *b*, of a dwarf *Golden Pippin-tree*, the same from which I cut the branch in Experiment 26: As the transverse cut *i* imbibed the water, the mercury rose 5 inches obliquely in the tube *z*, and 4 inches perpendicular.

In this, as also in many of the preceding Experiments there were several wounds in that part of the branch which was within the large tube *r i*; which were made by cutting off little lateral twigs, and swelling eyes, that the branch might easily enter the tube: And if these wounds (thro' which the air always issued plentifully) were well covered with sheeps-gut, bound over with packthread, it would in a good measure prevent the inconvenience: But I always found that my Experiments of this kind succeeded best, when that part of the branch which was to enter the tube *r i*, was clear of all knots or wounds; for when there were no knots, the liquor passed most freely, and less air issued out.

The same day I fixed in the same manner a gage to an *Apricot-tree*, it raised the mercury 3 inches; and tho' all the water was soon imbibed, yet the mercury rose every day an inch, for many days, and subsided at nights; so that the branch must daily imbibe thus much air, and remit it at night.

EXPERIMENT XXX.

We have a further proof of the influence of the leaves in raising the sap in this following Experiment.

August 6th, I cut off a large *Russet Pippin a* (Fig. 15.) with a stalk $1+\frac{1}{2}$ inch long, and 12 adjoyning leaves *g* growing to it.

I cemented the stalk fast into the upper end of the tube *d*, which tube was 6 inches long, and $\frac{1}{4}$ inch diameter; as the stalk imbibed the water, it raised the mercury to *z*, four inches high.

I fix'd another Apple of the same size and tree, in the same manner, but first pulled off the leaves; it raised the mercury but one inch; I fixed in the same manner a like bearing twig with 12 leaves on it, but no apple; it raised the mercury 3 inches.

I then took a like bearing twig, without either leaves or apple, it raised the mercury $\frac{1}{4}$ inch.

So a twig with an apple and leaves raised the mercury 4 inches, one with leaves only 3 inches, one with an apple without leaves 1 inch.

A Quince which had two leaves, just at the twig's insertion into it, raised the mercury $2+\frac{1}{2}$ inches and held it up a considerable time.

A sprig of Mint fix'd in the same manner, raised the mercury $3+\frac{1}{2}$ inch, equal to 4 feet + 5 inches height of water.

EXPERIMENT XXXI.

I tryed also the imbibing force of a great variety of trees, by fixing Aqueo-mercurial gages to branches of them cut off, as in Experiment 22.

The Pear, Quince, Cherry, Walnut, Peach, Apricot, Plumb, Black-thorns, White-thorns, Gooseberry, Water-Elder, Sycamore, raised the mercury from 6 to 3 inches high: Those which imbibed water most freely, in the Experiments of the first chapter, raised the mercury highest in these Experiments, except the Horse-Chesnut, which tho' it imbibed water most freely, yet raised the mercury but one inch, because the air passed very fast thro' its sap-vessels into the gage.

The following raised the mercury but 1 or 2 inches, *viz.* the Elm, Oak, Horse-Chesnut, Filberd, Fig, Mulberry, Willow, Sallow, Osier, Ash, Lynden, Currans.

The Ever-greens, and following trees and plants, did not raise it at all; the Laurel, Rosemary, Laurus-Tinus, Philarea, Fuz, Rue, Berberry, Jessamine, Cucumber-branch, Pumkin, Jerusalem Artichoke.

Experiment XXXII.

We have a further proof of the great force, with which vegetables imbibe moisture, in the following Experiment, *viz.* I filled near full with Pease and Water, the iron Pot (Fig. 37.) and laid on the Pease a leaden cover, between which, and the sides of the Pot, there was room for the air which came from the Pease, to pass freely. I then layed one hundred eighty four pounds weight on them, which (as the Pease dilated by imbibing the water) they lifted up. The dilatation of the Pease is always equal to the quantity of water they imbibe: For if a few Pease be put into a Vessel, and that Vessel be filled full of water, tho' the Pease dilate to near double their natural size; yet the water will not flow over the Vessel, or at most very inconsiderably, on account of the expansion of little air bubbles, which are issuing from the Pease. Being desirous to try whether they would raise a much greater weight, by means of a lever with weights at the end of it, I compressed several fresh parcels of Pease in the same Pot, with a force equal to 1600, 800, and 400 pounds; in which Experiments, tho' the Pease dilated, yet they did not raise the lever, because what they increased in bulk was, by the great incumbent weight, pressed into the interstices of the Pease, which they adequately filled up, being thereby formed into pretty regular Dodecahedrons.

We see in this Experiment the vast force with which swelling Pease expand, and 'tis doubtless a considerable part of the same force which is exerted, not only in pushing the Plume upwards into the air, but also in enabling the first shooting radicle of the Pea, and all its subsequent tender Fibres, to penetrate and shoot into the earth.

Experiment XXXIII.

We see, in the Experiments of this chapter, many instances of the great efficacy of attraction; that universal principle which is so operative in all the very different works of nature; and is most eminently so in vegetables, all whose minutest parts are curiously ranged in such order, as is best adapted by their united force, to attract proper nourishment.

And we shall find in the following Experiment, that the dissevered particles of vegetables, and of other bodies, have a strong attractive power when they lay confused.

That the particles of wood are specifically heavier than water (and can therefore strongly attract it) is evident, because several sorts of wood sink immediately; others (even cork) when their interstices are well soaked, and filled with water; others (as the Peruvian Bark) sink when very finely pulverized, because all their cavities which made them swim, are thereby destroyed.

In order to try the imbibing power of common wood ashes, I filled a glass tube *c r i*, 3 feet long, and $\frac{7}{8}$ of an inch diameter (Fig. 16.) with well dryed and sifted wood ashes; pressing them close with a rammer, I tyed a piece of linen over the end of the tube at *i*, to keep the ashes from falling out; I then cemented the tube *c* fast at *r* to the Aqueo-mercurial gage *r z*, and when I had filled the gage full of water, I immersed it in the cistern of mercury *x*; Then to the upper end of the tube *c*, at *o* I screwed on the mercurial gage *a b*.

The ashes as they imbibed the water drew the mercury up 3 or 4 inches in a few hours towards *z*; but the three following days it rose but 1 inch, $\frac{1}{2}$ inch, and $\frac{1}{4}$, and so less and less, so that in 5 or 6 days it ceased rising: The highest it rose was 7 inches which was equal to raising water 8 feet high.

This had very little effect on the mercury in the gage *a b*, unless it were, that it would rise a little, *viz.* an inch or little more in the gage at *a*, as it were by the suction of the ashes, to supply some of the air bubbles which were drawn out at *i*.

But when I separated the tube *c o* from the gage *r z*, and set the end *i* in water, then the moisture (being not restrained as before) rose faster and higher in the ashes *c o*, and depressed the mercury at *a*, so as to be 3 inches lower than in the leg *b*, by driving the air upwards, which was intermixed with the ashes.

I filled another tube 8 feet long, and $\frac{1}{2}$ inch diameter with red lead; and affixed it in the place of *c o* to the gages *a b, r z*. The mercury rose gradually 8 inches to *z*.

In both these Experiments, the end *i* was covered with innumerable air bubbles, many of which continually passed off, and were succeeded by others, as at the transverse cuts in the Experiments of this chapter. And as there, so in these, the quantity of air bubbles decreased every day, so as at last to have very few: The part *i* immersed in the water, being become so saturate therewith, as to leave no room for air to pass.

After 20 days I picked the minium out of the tube, and found the water had risen 3 feet 7 inches, and would no doubt have risen higher, if it had not been clogged by the mercury in the gage *z*. For which reason the moisture rose but 20 inches in the ashes, where it would otherwise have risen 30 to 40 inches,

And as Sir *Isaac Newton* (in his Opticks query 31.) observes, "The water rises up to this height, by the action only of those particles of the ashes which are upon the surface of the elevated water; the particles which are within the water, attracting or repelling it as much downwards as upwards; and therefore the action of the particles is very strong: But the particles of the ashes being not so dense and close together as those of glass, their action is not so strong as that of glass, which keeps quick-silver suspended to the height of 60 or 70 inches, and therefore acts with a force, which would keep water suspended to the height of above 60 feet.

"By the same principle, a sponge sucks in water, and the glands in the bodies of animals, according to their several natures and dispositions, suck in various juices from the blood."

And by the same principle it is, that we see in the preceding Experiments plants imbibe moisture so vigorously up their fine capillary vessels; which moisture, as it is carryed off in perspiration, (by the action of warmth,) thereby gives the sap vessels liberty to be almost continually attracting of fresh supplies, which they could not do, if they were full saturate with moisture: For without perspiration the sap must necessarily stagnate, notwithstanding the sap vessels are so curiously adapted by their exceeding fineness, to raise the sap to great heights, in a reciprocal proportion to their very minute diameters.

CHAP. III.

Experiments, shewing the force of the sap in the Vine in the bleeding season.

Having in the first chapter shewn many instances of the great quantities imbibed, and perspired by trees, and in the second chapter, seen the force with which they do imbibe moisture; I propose next to give an account of those Experiments, which prove with what great force the sap of the Vine is pushed forth, in the bleeding season.

EXPERIMENT XXXIV.

March 30th at 3 *p. m.* I cut off a *Vine* on a western aspect, within seven inches of the ground, the remaining stump *c* (Fig. 17) had no lateral branches: It was 4 or 5 years old, and $\frac{3}{4}$ inch diameter. I fix'd to the top of the stump by means of the brass collar *b*, the glass tube *b f*, seven feet long, and $\frac{1}{4}$ inch diameter; I secured the joynt *b* with stiff cement made of melted Beeswax and Turpentine, and bound it fast over with several folds of wet bladder and pack-thread: I then screwed a second tube *f g* to the first, and then a third *g a* to 25 feet height.

The stem not bleeding into the tube, I filled the tube two feet high with water, the water was imbibed by the stem within 3 inches of the bottom, by 8 a clock that evening. In the night it rained a small shower. The next morning at $6+\frac{1}{2}$ the water was risen three inches above what it was fallen to last night at eight a clock. The *Thermometer* which hung in my porch was 11 degrees above the freezing point,

March 31st from $6+\frac{1}{2}$ *a m*, to 10 *p. m.* the sap rose $8+\frac{1}{4}$ inches, *April* 1st at 6 *a m. Thermometer* 3 degrees above the freezing point, and a white hoar frost, the sap rose from ten a clock last night $3+\frac{1}{4}$ inches more; and so continued rising daily till it was above 21 feet high, and would very probably have risen higher, if the joynt *b* had not several times leaked: After stopping of which it would rise sometimes at the rate of an inch in 3 minutes, so as to rise 10 feet or more in a day. In the chief bleeding season it would continue rising night and day, but much more in the day than night, and most of all in the greatest heat of the day; and what little sinking it had of 2 or 3 inches was always after sunset, which I suspect was principally occasioned by the shrinking and contraction of the cement at *b*, as it grew cool.

When the sun shined hot upon the Vine, there was always a continued series of air bubbles, constantly ascending from the stem thro' the sap in the tube, in so great plenty as to make a large froth on the top of the sap, which shews the great quantity of air which is drawn in thro' the roots and stem.

From this Experiment we find a considerable energy in the root to push up sap in the bleeding season.

This put me upon trying, whether I could find any proof of such an energy, when the bleeding season was over, in order to which

Experiment XXXV.

July 4th at noon, I cut off within 3 inches of the ground, another *Vine* on a south aspect, and fixed to it a tube 7 feet high, as in the foregoing Experiment; I filled the tube with water, which was imbibed by the root the first day, at the rate of a foot in an hour, but the next day much more slowly, yet it was continually sinking, so that at noon day I could not see it so much as stationary.

Yet by Experiment the 3d, on the Vine in the garden pot, it is plain, that a very considerable quantity of sap was daily pressing thro' this stem, to supply the perspiration of the leaves, before I cut the vine off. And if this great quantity were carried up by pulsion or trusion, it must needs have risen out of the stem into the tube.

Now since this flow of sap ceases at once, as soon as the Vine was cut off the stem, the principal cause of its rise must at the same time be taken away, *viz.* the great perspiration of the leaves.

For tho' it is plain by many Experiments, that the sap enters the sap vessels of plants with much vigour, and is probably carried up to great heights in those vessels, by the vigorous undulations of the sun's warmth, which may reciprocally cause vibrations in the vesicles and sap vessels, and thereby make them dilate and contract a little; yet it seems as plain (from many Experiments, as particularly Exper. 13, 14, 15. and Exper. 43.) where tho' we are assured that a great quantity of water passed by the notch cut 2 or 3 feet above the end of the stem; yet was the notch very dry, because the attraction of the perspiring leaves was much greater than the force of trusion from the column of water. From these Experiments, I say, it seems evident, that the capillary sap vessels, out of the bleeding season, have little power to protrude sap in any plenty beyond their orifices; but as any sap is evaporated off, they can by their strong attraction (assisted by the genial warmth of the sun) supply the great quantities of sap drawn off by perspiration.

Experiment XXXVI.

April 6th at 9. *a m.* rain the evening before, I cut off a Vine on a southern aspect, at *a* (Fig. 18.) two feet nine inches from the ground, the remaining stem *a b*, had no lateral branches, it was $\frac{7}{8}$ inch diameter, I fixed on it the mercurial gage *a y*. At 11 *a. m.* the mercury was risen to *z*, 15 inches higher than in the leg *x*, being pushed down at *x*, by the force of the sap which came out of the stem at *a*.

At 4 *p. m.* it was sunk an inch in the leg *z y*.

April 7th at 8 *a. m.* risen very little, a fog: at 11 *a. m.* 'tis 17 inches high, and the fog gone.

April 10th at 7 *a. m.* mercury 18 inches high; I then added more mercury, so as to make the surface *z* 23 inches higher than *x*; the sap retreated very little into the stem, upon this additional weight, which shews with what an absolute force it advances: at noon it was sunk one inch.

April 11th at 7 *a. m.* $24+\frac{3}{4}$ inches high, sun-shine: at 7 *p. m.* 18 inches high.

April 14th at 7 *a. m.* $20+\frac{1}{4}$ inches high, at 9 *a. m.* $22+\frac{1}{2}$, fine warm sun-shine; here we see that the warm morning sun gives a fresh vigour to the sap. At 11 *a. m.* the same day $16+\frac{1}{2}$, the great perspiration of the stem makes it sink.

April 16th at 6 *a. m* $19+\frac{1}{2}$ rain. At 4 *p. m.* 13 inches. The sap (in the foregoing Experiment, numb. 34.) risen this day since noon 2 inches, while this sunk by the perspiration of the stem; which there was little room for, in the very short stem of the other.

April 17 at 11 *a. m.* $24+\frac{1}{4}$ inch high, rain and warm; at 7 *p. m.* $29+\frac{1}{2}$, fine warm rainy weather, which made the sap rise all day, there being little perspiration by reason of the rain.

April 18th at 7 *a. m.* $32+\frac{1}{2}$ inches high, and would have risen higher, if there had been more mercury in the gage; it being all forced into the leg *y z*. From this time to *May* 5th, the force gradually decreased.

The greatest height of the mercury being $32+\frac{1}{2}$ inches; the force of the sap was then equal to 36 feet $5+\frac{1}{3}$ inches height of water.

Here the force of the rising sap in the morning is plainly owing to the energy of the root and stem. In another like mercurial gage, (fixed near the bottom of a Vine which run

20 feet high) the mercury was raised by the force of the sap 38 inches equal to 43 feet + 3 inches + $\frac{1}{3}$ height of water.

Which force is near five times greater than the force of the blood in the great crural artery of a Horse; seven times greater than the force of the blood in the like artery of a Dog; and eight times greater than the blood's force in the same artery of a fallow Doe: Which different forces I found by tying those several animals down alive upon their backs; and then laying open the great left crural artery, where it first enters the thigh, I fixed to it (by means of two brass pipes, which run one into the other) a glass tube of above ten feet long, and $\frac{1}{8}$th of an inch diameter in bore: In which tube the blood of one Horse rose eight feet, three inches, and the blood of another Horse eight feet nine inches. The blood of a little Dog six feet and half high: In a large Spaniel seven feet high. The blood of the fallow Doe mounted five feet seven inches.

EXPERIMENT XXXVII.

April 4th, I fixed three mercurial gages (Fig. 19.) *a, b, c* to a *Vine*, on a south-east aspect, which was 50 feet long, from the root to the end *r u*. The top of the wall was 11 + $\frac{1}{2}$ feet high; from *i* to *k*, 8 feet; from *k* to *e*, 6 feet + $\frac{1}{2}$; from *e* to *a*, 1 foot 10 inches; from *e* to *o*, 7 feet; from *o* to *b*, 5 + $\frac{1}{2}$ feet; from *o* to *c*, 22 feet 9 inches; from *o* to *u*, 32 feet 9 inches.

The branches to which *a* and *c* were fixed were thriving shoots two years old, but the branch *o b* was much older.

When I first fixed them, the mercury was pushed by the force of the sap, in all the gages down the legs 4, 5, 13, so as to rise nine inches higher in the other legs.

The next morning at 7 *a. m.* the mercury in *a* was pushed 14 + $\frac{1}{4}$ inches high, in *b* 12 + $\frac{1}{4}$ in *c* 13 + $\frac{1}{2}$.

The greatest height to which they pushed the sap severally was *a* 21 inches, *b* 26 inches, *c* 26 inches.

The mercury constantly subsided by the retreat of the sap about 9 or 10 in the morning, when the Sun grew hot; but in a very moist foggy morning the sap was later before it retreated, *viz.* till noon, or some time after the fog was gone.

About 4 or 5 a clock in the afternoon, when the Sun went off the Vine, the sap began to push afresh into the gages, so as to make the mercury rise in the open legs; but it always rose fastest from Sunrise till 9 or 10 in the morning.

The sap in *b* (the oldest stem) plaid the most freely to and fro, and was therefore soonest affected with the changes from hot to cool, or from wet to dry, and *vice versâ.*

And *April* 10, toward the end of the bleeding season, *b* began first to suck up the mercury from 6 to 5, so as to be 4 inches higher in that leg than the other. But *April* 24, after a night's rain, *b* pushed the mercury 4 inches up the other leg, *a* did not begin to suck till *April* 29, *viz.* 9 days after *b*; *c* did not begin to suck till *May* 3. *viz.* 13 days after *b*, and 4 days after *a. May* 5. at 7 *a. m. a* pushed 1 inch, *c* $1+\frac{1}{2}$, but towards noon they all three sucked.

I have frequently observed the same difference in other Vines, where the like gages have been fixed at the same time, to old and young branches of the same Vine, *viz.* the oldest began first to suck.

In this Experiment we see the great force of the sap, at 44 feet 3 inches distance from the root, equal to the force of a column of water 30 feet 11 inches $+\frac{3}{4}$ high.

From this Experiment we see too, that this force is not from the root only, but must also proceed from some power, in the stem and branches: For the branch *b* was much sooner influenced by changes from warm to cool, or dry to wet, and *vice versâ*, than the other two branches *a* or *c*; and *b* was in an imbibing state, 9 days before *a*, which was all that time in a state of pushing sap; and *c* pushed 13 days after *b* had ceased pushing, and was in an imbibing state.

Which imbibing state Vines and Apple-trees continue in all the summer, in every branch, as I have found by fixing the like gages to them in *July*.

Experiment XXXVIII.

March 10, at the beginning of the bleeding season, (which is many days sooner or later, according to the coldness or warmth, moisture or dryness of the season) I then cut off a branch of *a vine b f c g* at *b*, (Fig. 20.) which was 3 or 4 years old, and cemented fast on it a brass-collar, with a screw in it; to that I screwed another brass collar, which was cemented fast to the glass tube *z*, 7 feet long and $\frac{1}{4}$ inch diam. (which I find to be the properest diam.) to that I screwed others, to 38 feet height. These tubes were fastened and secured in long wooden tubes, 3 inches square, one side of which was a door opening upon hinges; the use of those wooden tubes was to preserve the glass tubes from being broke by the freezing of the sap in them in the night. But when the danger of hard frosts was pretty well over, as at the beginning of *April*, then I usually fix'd the glasses without the wooden tubes, fastening them to scaffold poles, or two long iron spikes drove into the wall.

Before I proceed to give an account of the rise and fall of the sap, in the tubes I will first describe the manner of cementing on the brass collar *b*, to the stem of the Vine in which I have been often disappointed, and have met with difficulties; it must therefore be done with great care.

Where I design to cut the stem, I first pick off all the rough stringy bark carefully with my nails to avoid making any wound thro' the green inner bark; then I cut off the branch at *i*, (Fig. 21.) and immediately draw over the stem a piece of dried sheeps-gut, which I tye fast, as near the end of the stem as I can, so that no sap can get by it; the sap being confined in the gut *i f*: Then I wipe the stem at *i* very dry with a warm cloth, and tie round the stem a stiff paper funnel *x i*, binding it fast at *x* to the stem; and pinning close the folds of the paper from *x* to *i*: Then I slide the brass collar *r* over the gut, and immediately pour into the paper funnel melted brickdust cement, and then set the brass-collar into it; which collar is warmed, and dipped before in the cement, that it may the better now adhere: When the cement is cold, I pull away the gut, and screw on the glass tubes.

But finding some inconvenience in this hot cement (because its heat kills the sap vessels near the bark, as is evident by their being discoloured) I have since made use of the cold cement of Bees-wax and Turpentine, binding it fast over with wet bladder and pack-thread, as in *Exper.* 34.

Instead of brass-collars, which screwed into each other, I often (especially with the Syphons in Exper. 37, and 38.) made use of two brass collars, which were turned a little tapering, so that one entered and exactly fitted the other.

This joining of the two collars was effectually secured from leaking, by first anointing them with a soft cement; and they were secured from being disjoined, by the force of the ascending sap, by twisting pack-thread round the protuberant knobs on the sides of the collars. When I would separate the collars, I found it necessary (except in hot Sunshine) to melt the soft cement by applying hot irons on the outside of the collars.

It is needful to shade all the cemented joints from the Sun with loose folds of paper, else its heat, will often melt them, and so dilate the cement, as to make it be drove forcibly up the tube, which defeats the Experiment.

The *Vines* to which the tubes in this Experiment were fixed, were 20 feet high from the roots to their top; and the glass tubes fixed at several heights *b* from the ground, from six to two feet.

The sap would rise in the tube the first day, according to the different vigor of the bleeding state of the Vine, either 1, 2, 5, 12, 15, or 25 feet; but when it had got to its greatest

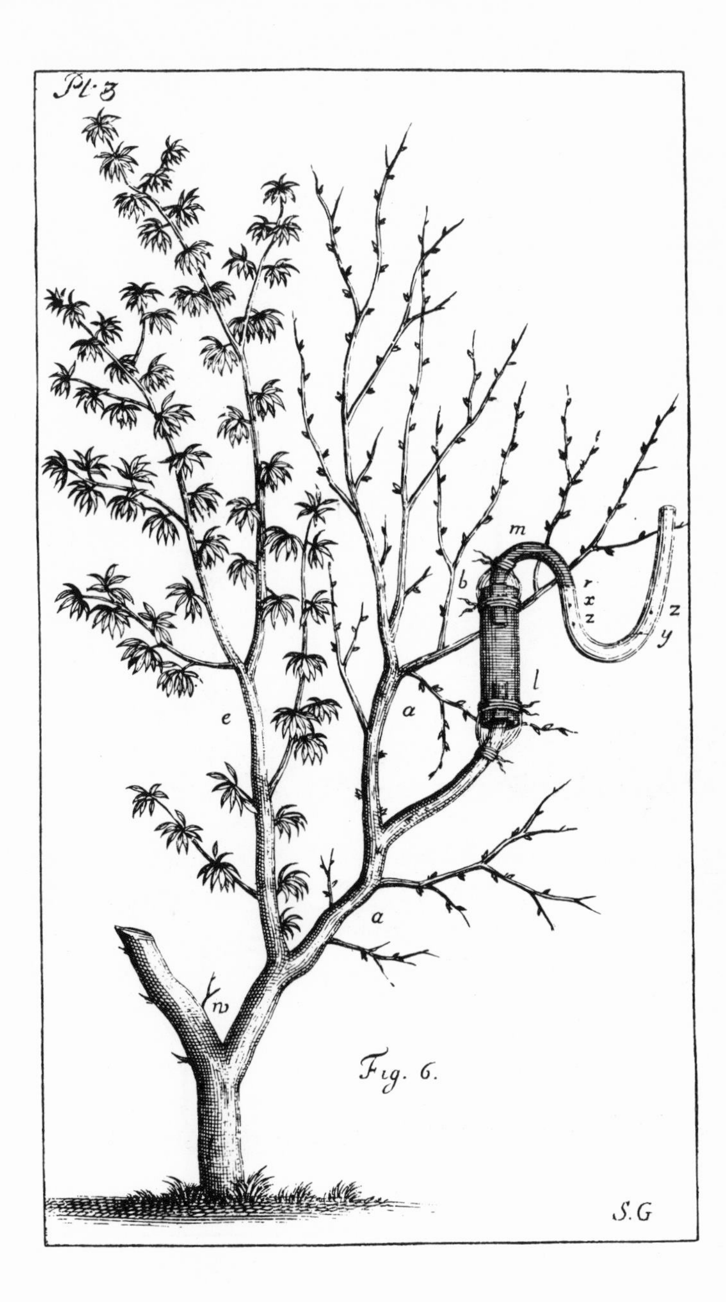

Fig. 6.

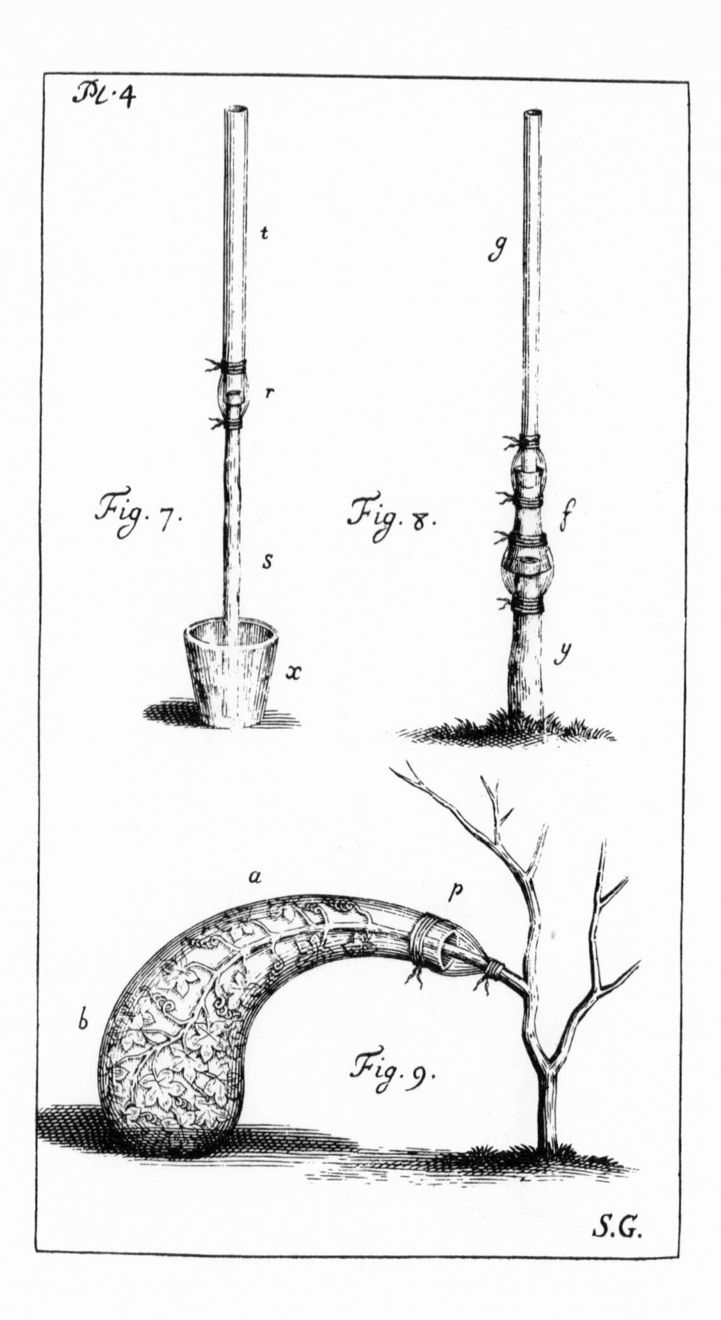
Pl. 4
t
r
Fig. 7.
s
x
g
Fig. 8.
f
y
a
p
b
Fig. 9.
S.G.

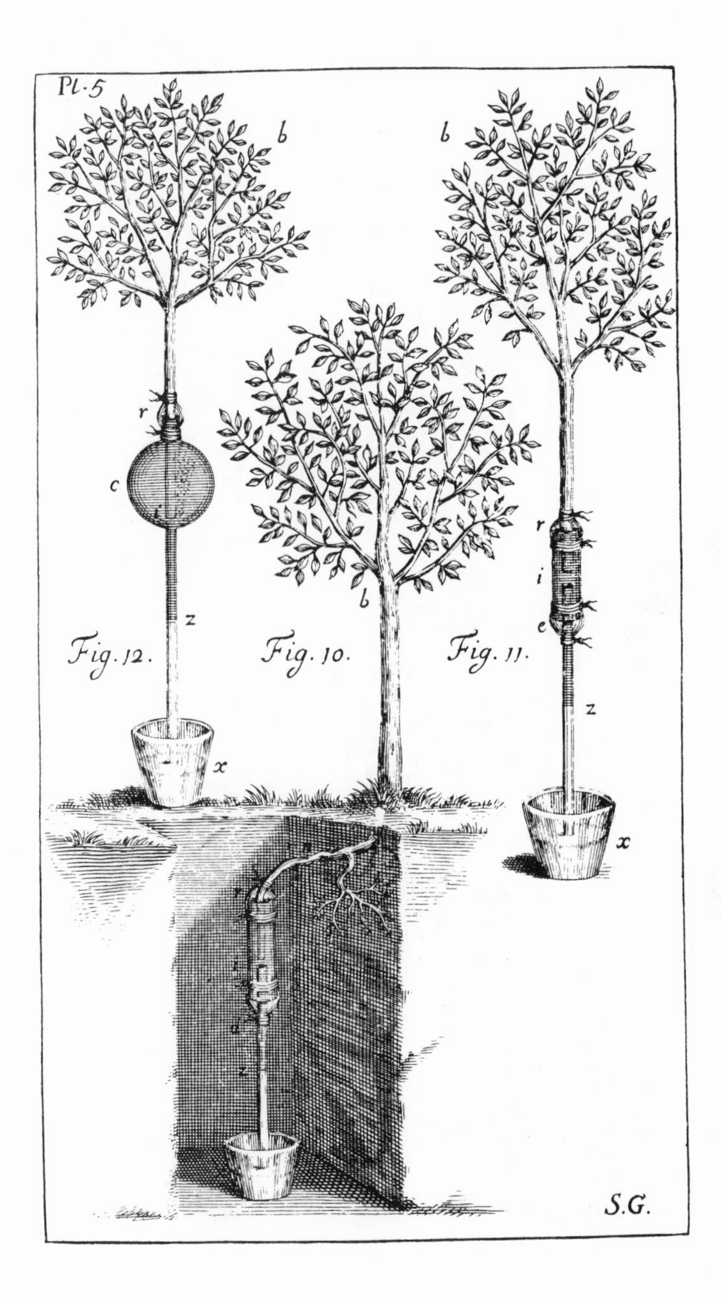
Pl. 5
b
b
b
r
c
z
Fig. 12.
Fig. 10.
Fig. 11.
r
i
e
z
x
x
z
S.G.

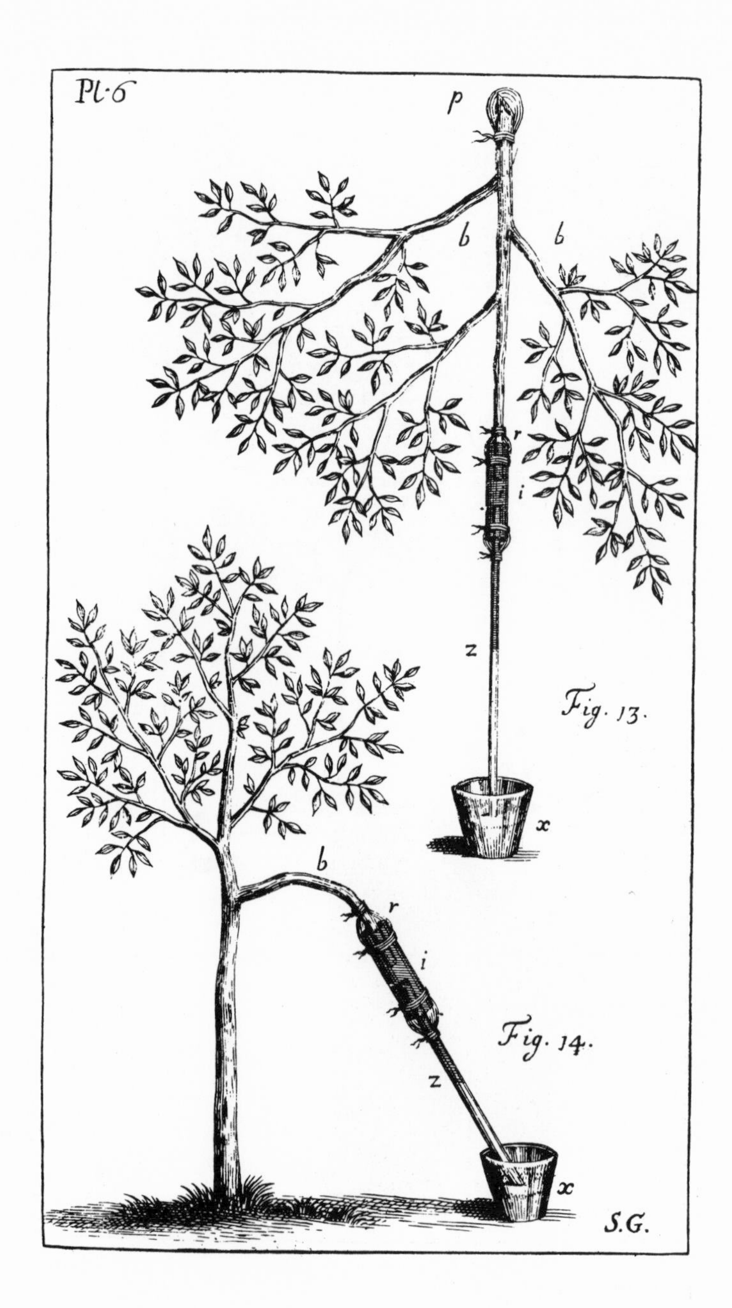
Pl·6
p
b
b
r
i
z
Fig. 13.
x
b
r
i
Fig. 14.
z
x
S.G.

height for that day, if it was in the morning, it would constantly begin to subside towards noon.

If the weather was very cool about the middle of the day, it would subside only from 11 or 12, to two in the afternoon; but if it were very hot weather, the sap would begin to subside at 9 or 10 a clock, and continue subsiding till 4, 5, or 6 in the evening, and from that time it would continue stationary for an hour or two; after which it would begin to rise a little, but not much in the night, nor till after the sun was up in the morning, at which time it rose fastest.

The fresher the cut of the Vine was, and the warmer the weather, the more the sap would rise, and subside in a day, as 4 or 6 feet.

But if it were 5 or 6 days since the Vine was cut, it would rise or subside but little; the sap-vessels at the transverse cut being saturate and contracted.

But if I cut off a joint or two off the stem, and new fixed the tube, the sap would then rise and subside vigorously.

Moisture and warmth made the sap most vigorous.

If the beginning or middle of the bleeding season being very kindly, had made the motion of the sap vigorous; that vigour would immediately be greatly abated by cold easterly winds.

If in the morning, while the sap is in a rising state, there was a cold wind with a mixture of sun-shine and cloud; when the Sun was clouded, the sap would immediately visibly subside, at the rate of an inch in a minute for several inches, if the Sun continued so long clouded: But as soon as the Sun-beams broke out again, the sap would immediately return to its then rising state, just as any liquor in a *Thermometer* rises and falls with the alternacies of heat and cold; whence 'tis probable, that the plentiful rise of the sap in the Vine in the bleeding season, is effected in the same manner.

When three Tubes were fixed at the same time to Vines on an eastern, a southern, and a western Aspect, round my

Porch, the sap would begin to rise in the morning first in the eastern-tube, next in the southern, and last in the western-tube: And towards noon it would accordingly begin to subside, first in the eastern-tube, next in the southern, and last in the western-tube.

Where two branches arose from the same old western trunk, 15 inches from the ground; and one of these branches was spread on a southern, and the other on a western Aspect; and glass tubes were at the same time fixed to each of them; the sap would in the morning, as the Sun came on, rise first in the southern, then in the western-tube; and would begin to subside, first in the southern, then in the western-tube.

Rain and warmth, after cold and dry, would make the sap rise all the next day, without subsiding, tho' it would rise then slowest about noon; because in this case the quantity imbibed by the root, and raised from it, exceeded the quantity perspired.

The sap begins to rise sooner in the morning in cool weather, than after hot days; the reason of which may be, because in hot weather much being evaporated, it is not so soon supplied by the roots as in cool weather, when less is evaporated.

In a prime bleeding season I fixt a tube 25 feet long to a thriving branch two years old, and two feet from the ground, where it was cut off; the sap flowed so briskly, as in two hours to flow over the top of the Tube, which was 7 feet above the top of the Vine; and doubtless would have risen higher, if I had been prepared to lengthen the tube.

When at the distance of 4 or 5 days, tubes were affixed to two different branches, which came from the same stem, the sap would rise highest in that which was last fixed; yet if in the fixing the second tube there was much sap lost, the sap would subside in the first tube; but they would not afterwards have their sap in Equilibrio; *i. e.* the surface of the sap in each was at very unequal heights; the reason of which is, because of the difficulty with which the sap passes thro' the almost saturate and contracted Capillaries of the first cut stem.

In very hot weather many air bubbles would rise, so as to make a froth an inch deep, on the top of the sap in the tube.

I fixt a small air Pump to the top of a long Tube, which had 12 feet height of sap in it; when I pumped, great plenty of bubbles arose, tho' the sap did not rise, but fall a little, after I had done pumping.

In Experiment 34. (where a Tube was fixed to a very short stump of a Vine, without any lateral branches) we find the sap rose all day, and fastest of all in the greatest heat of the day: But by many observations under the 37th and this 38th Experiments, we find the sap in the tubes constantly subsided as the warmth came on towards the middle of the day, and fastest in the greatest heat of the day. Whence we may reasonably conclude, (considering the great perspirations of trees, shewn in the first chapter) that the fall of the sap, in these sap gages, in the middle of the day, especially in the warmer days, is owing to the then greater perspiration of the branches, which perspiration decreases, as the heat decreases towards evening, and probably wholly ceases when the dews fall.

But when towards the latter end of *April*, the spring advances, and many young shoots are come forth, and the surface of the Vine is greatly increased and enlarged, by the expansion of several leaves; whereby the perspiration is much increased, and the sap more plentifully exhausted, it then ceases to flow in a visible manner, till the return of the following spring.

And as in the Vine, so is the case the same in all the bleeding trees, which cease bleeding as soon as the young leaves began to expand enough, to perspire plentifully, and to draw off the redundant sap. Thus the bark of Oaks and many other trees most easily separates while it is lubricated with plenty of sap: But as soon as the leaves expand sufficiently to perspire off plenty of sap, the bark will then no longer run (as they term it) but adheres most firmly to the wood.

Experiment XXXIX.

In order to try if I could perceive the stem of the Vine dilate and contract with heat or cold, wet or dry, a bleeding or not bleeding season, some time in *February*, I fixt to the stem of a Vine an instrument in such a manner, that if the stem had dilated or contracted but the one hundredth part of an inch, it would have made the end of the instrument, (which was a piece of strong brass-wire, eighteen inches long) rise or fall very sensibly about one tenth of an inch; but I could not perceive the instrument to move, either by heat or cold, a bleeding or not bleeding season. Yet whenever it rained the stem dilated so as to raise the end of the instrument or lever $\frac{3}{10}$ of an inch, and when the stem was dry it subsided as much.

This Experiment shews, that the sap (even in the bleeding season) is confined in its proper vessels, and that it does not confusedly pervade every interstice of the stem, as the rain does, which entering at the perspiring pores, soaks into the interstices, and thereby dilates the stem.

CHAP. IV.

Experiments, shewing the ready lateral motion of the sap, and consequently the lateral communication of the sap vessels. The free passage of it from the small branches towards the stem, as well as from the stem to the branches. With an account of some Experiments, relating to the circulation or non-circulation of the sap.

EXPERIMENT XL.

In order to find whether there was any lateral communication of the sap and sap vessels, as there is of the blood in animals, by means of the ramifications, and lateral communications of their vessels:

August 15th, I took a young *Oak branch* $\frac{7}{8}$ inches diameter, at its transverse cut, 6 feet high, and full of leaves. Seven inches from the bottom, I cut a large gap to the pith, an inch long, and of an equal depth the whole length: And 4 inches above that, on the opposite side, I cut such another gap; I set the great end of the stem in water: It imbibed and perspired in two nights and two days 13 ounces, while another like oak branch, somewhat bigger than this, but with no notch cut in its stem, imbibed 25 ounces of water.

At the same time I tryed the like Exper. with a *Duke-Cherry branch*; it imbibed and perspired 23 ounces in 9 hours the first day, and the next day 15 ounces.

At the same time I took another *Duke-Cherry branch*, and cut 4 such square gaps to the pith, 4 inches above each other; the 1st *North*, 2d *East*, 3d *South*, 4th *West*: It had a long slender stem, four feet length, without any branches, only at the very top; yet it imbibed in 7 hours day 9 ounces, and in two days and two nights 24 ounces.

We see in these Experiments a most free lateral communication of the sap and sap vessels, these great quantities of liquor having passed laterally by the gaps; for by Experiment 13, 14, 15. (on Cylinders of wood) little evaporated at the gaps.

And in order to try whether it would not be the same in branches as they grew on trees, I cut two such opposite gaps in a *Duke-Cherry branch*, 3 inches distant from each other: The leaves of this branch continued green, within 8 or 10 days, as long as the leaves on the other branches of the same tree.

The same day, *viz. August* 15th, I cut two such opposite gaps 4 inches distant, in an horizontal young thriving *Oak-branch*; it was 1 inch diameter, 18 days after many of the leaves begun to turn yellow, which none of the leaves of other boughs did then.

The same day I cut off the bark for one inch length, quite round a like branch of the same *Oak*; 18 days after the leaves were as green as any on the same tree; but the leaves fell off this and the foregoing branch early in the winter; yet continued on all the rest of the boughs of the tree (except the top ones) all the winter.

The same day I cut four such gaps, 2 inches wide, and 9 inches distant from each other, in the upright arm of a *Golden-Renate-tree*; the diameter of the branch was $2+\frac{1}{2}$ inch, the gaps faced the four cardinal points of the compass; the apples and leaves on this branch flourished as well as those on other branches of the same tree.

Here again we see the very free lateral passage of the sap, where the direct passage is several times intercepted.

Experiment XLI.

August 13. At noon I took a large branch of an *Apple-tree*, (Fig. 22.) and cemented up the transverse cut, at the great end *x*, and tyed a wet bladder over it: I then cut off the main top branch at *b*; where it was $\frac{6}{8}$ inch diameter, and set it thus inverted into the bottle of water *b*.

In three days and two nights it imbibed and perspired 4 pounds 2 ounces + $\frac{1}{2}$ of water, and the leaves continued green; the leaves of a bough cut off the same tree at the same time with this, and not set in water, had been withered 40 hours before. This, as well as the great quantities imbibed and perspired, shews, that the water was drawn from *b* most freely to *e*, *f*, *g*, *h*, and from thence down their respective branches, and so perspired off by the leaves.

This Experiment may serve to explain the reason, why the branch *b*, (Fig. 23.) which grows out of the root *c x*, thrives very well, notwithstanding the root *c x* is here supposed to be cut off at *c*, and to be out of the ground: For by many Experiments in the first and second chapters, it is evident, that the branch *b* attracts sap at *x* with great force: And by this present Experiment, 'tis as evident, that sap will be drawn as freely downwards from the tree to *x*, as from *c* to *x*, in case the end *c* of the root were in the ground; whence 'tis no wonder, that the branch *b* thrives well, tho' there be no circulation of the sap.

This Experiment 41, and Experiment 26, do also shew the reason why, where three trees (Fig. 24.) are inarched, and thereby incorporated at *x* and *z*, the middle tree will then grow, tho' it be cut off from its roots, or the root be dug out of the ground, and suspended in the air; *viz.* because the middle tree *b* attracts nourishment strongly at *x* and *z*, from the adjoyning trees *a c*, in the same manner as we see the inverted boughs imbibed water in these Exper. 26, and 41.

And from the same reason it is that Elders, Sallows, Willows, Briars, Vines, and most Shrubs, will grow in an inverted state, with their tops downwards in the earth.

Experiment XLII.

July 27th, I repeated Monsieur *Perault*'s Experiment, *viz.* I took *Duke-Cherry, Apple* and *Curran-Boughs*, with two branches each, one of which *a c* (Fig. 25.) I immersed in the large vessel of water *e d*, the other branch hanging in the open air: I hung on a rail, at the same time, other branches of the same sorts, which were then cut off. After three days, those on the rails were very much withered and dead, but the branches *b* were very green; in 8 days the branch *b* of the *Duke-Cherry* was much withered; but the *Currans* and *Apple-branch b* did not fade till the eleventh day: Whence 'tis plain, by the quantities that must be perspired in eleven days, to keep the leaves *b* green so long, and by the waste of the water, out of the vessel, that these boughs *b* must have drawn much water, from and through the other boughs and leaves *c*, which were immersed in the vessel of water.

I repeated the like Experiment on the branches of Vines and Apple-trees, by running their boughs as they grew into large glass chymical retorts full of water, where the leaves continued green for several weeks, and imbibed considerable quantities of water.

This shews how very probable it is, that rain and dew is imbibed by vegetables, especially in dry seasons.

Which is further confirmed by Experiments lately made on new-planted trees; where by frequently washing the bodies of the most unpromising, they have out-stripped the other trees of the same plantation. And Mr. *Miller* advises, "Now and then in an evening to water the head, and with a brush to wash and supple the bark all round the trunk, which (says he) I have often found very serviceable." *Supplement* to his *Gardener's Dictionary*, Vol. II under Planting.

EXPERIMENT XLIII.

August 20th, at 1 *p. m.* I took an *Apple-branch b*, (Fig. 26.) nine feet long, $1+\frac{3}{4}$ inch diameter, with proportional lateral branches, I cemented it fast to the tube *a*, by means of the leaden Syphon *l*: but first I cut away the bark, and last year's ringlet of wood, for 3 inches length to *r*. I then filled the tube with water, which was 12 feet long, and $\frac{1}{2}$ inch diameter, having first cut a gap at *y* through the bark, and last year's wood, 12 inches from the lower end of the stem: the water was very freely imbibed, *viz.* at the rate of $3+\frac{1}{2}$ inches in a minute. In half an hour's time I could plainly perceive the lower part of the gap *y* to be moister than before; when at the same time the upper part of the wound looked white and dry.

Now in this case the water must necessarily ascend from the tube, thro' the innermost wood, because the last year's wood was cut away, for 3 inches length all round the stem; and consequently, if the sap in its natural course descended by the last year's ringlet of wood, and between that and the bark (as many have thought) the water should have descended by the last year's wood, or the bark, and so have first moistened the upper part of the gap *y*; but on the contrary, the lower part was moisten'd, and not the upper part.

I repeated this Experiment with a large *Duke-Cherry branch*, but could not perceive more moisture at the upper, than the lower part of the gap, which ought to have been, if the sap descends by the last year's wood or the bark.

It was the same in a *Quince-branch* as the *Duke-Cherry*.

N. B. When I cut a notch in either of these branches, 3 feet above *r*, at *q*, I could neither see nor feel any moisture, notwithstanding there was at the same time a great quantity of water passing by; for the branch imbibed at the rate of 4, 3, or 2 inches *per* minute, of a column of water which was half inch diameter.

The reason of which dryness of the notch *q* is evident from Experiment 11, *viz.* because the upper part of the branch above the notch imbibed and perspired 3 or 4 times more water, than a column of 7 feet height of water in the tube could impel from the bottom of the stem to *q*, which was 3 feet length of stem; and consequently, the notch must necessarily be dry, notwithstanding so large a stream of water was passing by; *viz.* because the branch and stem above the notch was in a strongly imbibing state, in order to supply the great perspiration of the leaves.

Experiment XLIV.

August 9th at 10 *a. m.* I fix'd in the same manner (as in the foregoing Experiment) a *Duke-Cherry branch* 5 feet high, and 1 inch diameter, but did not cut away any of the bark or wood at the great end; I filled the tube with water, and then cut a slice off the bark an inch long, 3 inches above the great end; it bled at the lower part most freely, while the upper part continued dry.

The same day I tryed the same Experiment on an *Apple-branch*, and it had the same effect.

From these Experiments 'tis probable that the sap ascends between the bark and wood, as well as by other parts.

And since by other Experiments it is found that the greatest part of the sap is raised by the warmth of the Sun on the leaves, which seem to be made broad and thin for that purpose; for the same reason, it's most probable, it should rise also in those parts which are most exposed to the Sun, as the bark is.

And when we consider, that the sap vessels are so very fine, as to reduce the sap almost to a vapour, before it can enter them, the Sun's warmth on the bark should most easily dispose such rarified sap to ascend, instead of descending.

EXPERIMENT XLV.

July 27th, I took several *branches* of *Currans, Vines, Cherry, Apple, Pear* and *Plumb-tree*, and set the great ends of each in vessels of water *x*, (Fig. 31.) but first took the bark for an inch off one of the branches, as at *z*, to try whether the leaves above *z* at *b* would continue green longer than the leaves of any of the other branches *a*, *c*, *d*; but I could find no difference, the leaves withering all at the same time: Now if the return of the sap was stopped at *z*, then it would be expected, that the leaves at *b* should continue green longer than those on the other branches; which did not happen, neither was there any moisture at *z*.

EXPERIMENT XLVI.

In *August*, I cut off the bark for an inch round, of a young thriving Oak-branch on the North-West side of the tree. The leaves of this and another branch, which had the bark cut at the same time, fell early, *viz.* about the latter end of *October*, when the leaves of all the other branches of the same tree, except those at the very top of the tree, continued on all the winter.

This is a further proof, that less sap goes to branches which have the bark cut off, than to others.

The 19th of *April* following, the buds of this branch were 5 or 7 days forwarder than those of other branches of the same tree; the reason of which may probably be, because less fresh crude sap coming to this branch than the others, and the perspirations in all branches being *cæteris paribus* nearly equal, the lesser quantity of sap in this branch must sooner be inspissated into a glutinous substance, fit for new productions, than the sap of other branches, that abounded with a greater plenty of fresh thin sap.

The same is the reason why Apples, Pears, and many other fruits, which have some of their great sap vessels eaten asunder by insects bred in them, are ripe many days before the rest of the fruit on the same trees; as also that fruit which is gathered some time before it is ripe, will ripen sooner than if it had hung on the tree, tho' it will not be so good; because in these cases the worm-eaten fruit is deprived of part of its nourishment, and the green gathered fruit of all.

And for the same reason some fruits are sooner ripe towards the tops of the trees, than the other fruit on the same tree; *viz.* not only because they are more exposed to the sun; but also, because being at a greater distance from the root, they have somewhat less nourishment.

And this is, doubtless, one reason why plants and fruits are forwarder in dry, sandy or gravelly soils, than in moister soils; *viz.* not only, because those soils are warmer on account of their dryness; but also, because less plenty of moisture is conveyed up the plants; which plenty of moisture, tho' it promotes their growth, yet retards their coming to maturity. And for the same reason, the uncovering the roots of trees for some time, will make the fruit be considerably the forwarder.

And on the other hand, where trees abound with too great a plenty of fresh drawn sap, as is the case of trees whose roots are planted too deep in cold moist earth, as also of too luxuriant Peach and other Wall trees; or which comes almost to the same, where the sap cannot be perspired off in a due proportion; as in Orchards where trees stand too near each other, so as to hinder perspiration, whereby the sap is kept in too thin and crude a state; in all these cases little or no fruit is produced.

Hence also, in moderately dry summers, *cæteris paribus*, there is usually greatest plenty of fruit; because the sap in the bearing twigs and buds is more digested, and brought to a better consistence, for shooting out with vigour and firmness, than it is in cool moist summers: And this observation has been verified in the years 1723, 1724, and 1725. See an account of them under it. *Exp.* 20.

But to return to the subject of the motion of the sap; when the sap has first passed thro' that thick and fine strainer, the bark of the root, we then find it in greatest quantities, in the most lax part, between the bark and wood, and *that* the same thro' the whole tree. And if in the early spring, the Oak and several other trees were to be examined near the top and bottom, when the sap first begins to move, so as to make the bark easily run, or peel off, I believe it would be found that the lower bark is first moistened; whereas the bark of the top branches ought first to be moistened, if the sap descends by the bark: As to the Vine, I am pretty well assured that the lower bark is first moistened.

We see in many of the foregoing Experiments, what quantities of moisture trees do daily imbibe and perspire: Now the celerity of the sap must be very great, if that quantity of moisture must, most of it, ascend to the top of the tree, then descend, and ascend again, before it is carried off by perspiration.

The defect of a circulation in vegetables seems in some measure to be supplied by the much greater quantity of liquor, which the vegetable takes in, than the animal whereby its motion is accelerated; for by *Experiment* 1st, we find the Sunflower, bulk for bulk, imbibes and perspires 17 times more fresh liquor than a man every 24 hours.

Besides, nature's great aim in vegetables being only that the vegetable life be carried on and maintained, there was no occasion to give its sap the rapid motion, which was necessary for the blood of animals.

In animals, it is the heart which sets the blood in motion, and makes it continually circulate; but in vegetables we can discover no other cause of the sap's motion, but the strong attraction of the capillary sap vessels, assisted by the brisk undulations and vibrations, caused by the sun's warmth, whereby the sap is carried up to the top of the tallest trees, and is there perspired off thro' the leaves: But when the surface of the tree is greatly diminished by the loss of its leaves, then also the perspiration and motion of the sap is proportionably diminished, as is plain from many of the foregoing Experiments: So that the ascending velocity of the sap is principally accelerated by the plentiful perspiration of the leaves, thereby making room for the fine capillary vessels to exert their vastly attracting power, which perspiration is effected by the brisk rarifying vibrations of warmth: A power that does not seem to be any ways well adapted, to make the sap descend from the tops of vegetables by different vessels to the root.

If the sap circulated, it must needs have been seen descending from the upper part of large gashes, cut in branches, set in water, and with columns of water pressing on their bottoms in long glass tubes, in *Exp.* 43, and 44. In both which cases, it is certain that great quantities of water passed thro' the stem, so that it must needs have been seen descending, if the return of the sap downwards were by trusion or pulsion, whereby the blood in animals is returned thro' the veins to the heart: And that pulsion, if there were any, must necessarily be exerted with prodigious force, to be able to drive the sap thro' the finer capillaries. So that if there be a return of the sap downwards, it must be by attraction, and that a very powerful one, as we may see by many of these Experiments, and particularly by Experiment 11. But it is hard to conceive, what and where that power is which can be equivalent to that provision nature has made for the ascent of the sap in consequence of the great perspiration of the leaves.

The instances of the Jessamine tree, and of the Passion tree, have been looked upon as strong proofs of the circulation of the sap, because their branches, which were far below the inoculated Bud, were gilded: But we have many visible proofs in the Vine, and other bleeding trees, of the saps' receding back, and pushing forwards alternately, at different times of the day and night. And there is great reason to think, that the sap of all other trees has such an alternate, receding and progressive motion, occasioned by the alternacies of day and night, warm and cool, moist and dry.

For the sap in all vegetables does probably recede in some measure from the tops of branches, as the Sun leaves them; because its rarifying power then ceasing, the greatly rarified sap, and air mixt with it, will condense and take up less room than they did, and the dew and rain will then be strongly imbibed by the leaves, as is probable from Exper. 42. and several others; whereby the body and branches of the vegetable which have been much exhausted by the great evaporation of the day, may at night imbibe sap and dew from the leaves; for by several Experiments in the first chapter, plants were found to increase considerably in weight, in dewy and moist nights. And by other Experiments on the Vine in the third chapter, it was found, that the trunk and branches of Vines were always in an imbibing state, caused by the great perspiration of the leaves, except in the bleeding season; but when at night that perspiring power ceases, then the contrary imbibing power will prevail and draw the sap and dew from the leaves, as well as moisture from the roots.

And we have a further proof of this in Experiment 12, where by fixing mercurial gages to the stems of several trees, which do not bleed, it is found, that they are always in a strongly imbibing state, by drawing up the mercury several inches: Whence it is easy to conceive, how some of the particles of the gilded Bud, in the inoculated Jessamine, may be absorbed by it, and thereby communicate their gilding Miasma to the sap of other branches; especially when some months after the inoculation, the stock of the inoculated Jessamine is cut off a little above the Bud; whereby the stock, which was the counter acting part to the stem, being taken away, the stem attracts more vigorously from the Bud.

Another argument for the circulation of the sap, is, that some sorts of graffs will infect and canker the stocks they are grafted on: But by Exper. 12 and 37, where mercurial gages were fixed to fresh cut stems of trees, it is evident that those stems were in a strongly imbibing state; and consequently the cankered stocks might very likely draw sap from the graff, as well as the graff alternately from the stock; just in the same manner as leaves and branches do from each other, in the vicissitudes of day and night. And this imbibing power of the stock is so great, where only some of the branches of a tree are grafted, that the remaining branches of the stock will, by their strong attraction, starve those graffs; for which reason it is usual to cut off the greatest part of the branches of the stock, leaving only a few small ones to draw up the sap.

The instance of the Ilex grafted upon the *English* Oak, seems to afford a very considerable argument against a circulation. For if there were a free uniform circulation of the sap thro' the Oak and Ilex, why should the leaves of the Oak fall in winter, and not those of the Ilex?

Another argument, against a uniform circulation of the sap in trees as in animals, may be drawn from Exper. 37. where it was found by the three mercurial gages fixt to the same Vine, that while some of its branches changed their state of protruding sap into a state of imbibing, others continued protruding sap, one nine, and the other thirteen days longer.

In the second Vol. of Mr. *Lowthorp*'s *Abridgment of the Philos. Transac. p.* 708, is recited an Experiment of Mr. *Brotherton*'s, *viz.* A young Hazel *n*, Fig. 27., was cut into the body at *x z* with a deep gash; the parts of the body below at *z*, and above at *x*, were cleft upwards and downwards, and the splinters *x z* by wedges were kept off from touching each other, or the rest of the body. The following year, the upper splinter *x* was grown very much, but the lower splinter *z* did not grow, but the rest of the body grew, as if there had been no gash made: I have not yet succeeded in making this Experiment, the wind having broken at *x z* all the trees I prepared for it: But if there was a Bud at *x* which shot out leaves and none at *z*, then by Experiment 41. 'tis plain that those leaves might draw much nourishment thro' *t x*, and thereby make it grow; and I believe, if, *vice versâ*, there were a leaf bearing Bud at *z*, and none at *x*, that then the splinter *z* would grow more than *x*.

The reason of my conjecture, I ground upon this Experiment, *viz*. I chose two thriving shoots of a dwarf *Pear-tree l l a a*. Fig. 28, 29. At three quarters of an inch distance I took half an inch breadth of bark off each of them, in several places, *viz*. 2, 4, 6, 8, and at 10, 12, 14. every one of the remaining ringlets of bark had a leaf bearing bud, which produced leaves the following summer, except the ringlet 13, which had no such Bud. The ringlet 9 and 11 of *a a* grew and swelled at their bottoms, till *August*, but the ringlet 13 did not increase at all, and in *August* the whole shoot *a a* withered and dyed; but the shoot *l l* lives and thrives well, each of its ringlets swelling much at the bottom: Which swellings at their bottoms must be attributed to some other cause than the stoppage of the sap in its return downwards, because in the shoot *l l*, its return downwards is intercepted three several times by cutting away the bark at 2, 4, 6. The larger and more thriving the leaf bearing Bud was, and the more leaves it had on it, so much the more did the adjoining bark swell at the bottom.

Fig. 30. Represents the profile of one of the divisions in Fig. 28. split in halves, in which may be seen the manner of the growth of the last year's ringlet of wood shooting a little upwards at *x x*; and shooting downwards and swelling much more at *z z*; where we may observe, that what is shot endways, is plainly parted from the wood of the preceding year, by the narrow interstices *x r, z r*, whence it should seem, that the growth, of the yearly new ringlets of wood consists in the shooting of their fibres lengthways under the bark.

That the sap does not descend between the bark and the wood, as the favourers of a circulation suppose, seems evident from hence, *viz.* that if the bark be taken off for 3 or 4 inches breadth quite round, the bleeding of tree above that bared place will much abate, which ought to have the contrary effect, by intercepting the course of the refluent sap, if the sap descended by the bark.

But the reason of the abatement of the bleeding in this case may well be accounted for, from the manifest proof we have in these Experiments, that the sap is strongly attracted upwards by the vigorous operation of the perspiring leaves, and attracting Capillaries: But when the bark is cut off for some breadth below the bleeding place, then the sap which is between the bark and the wood below that disbarked place, is deprived of the strong attracting power of the leaves, *&c.* and consequently the bleeding wound cannot be supplied so fast with sap, as it was before the bark was taken off.

Hence also we have a hint for a probable conjecture why in the alternately disbarked sticks, *l l a a* Fig. 28, 29. the bark swelled more at the upper part of the disbarked places than at the lower, *viz.* because those lower parts were thereby deprived of the plenty of nourishment which was brought to the upper parts of those disbarked places by the strong attraction of the leaves on the Buds 7, *&c.* of which we have a further confirmation in the ringlet of bark, N°. 13. Fig. 29. which ringlet did not swell or grow at either end, being not only deprived of the attraction of the superior leaves, by the bared place N°. 12. but also without any leaf Bud of its own, whose branching sap Vessels, being like those of other leaf Buds rooted downwards in the wood, might thence draw sap, for the nourishment of its self and the adjoining bark N°. 13. But had these rooting sap vessels run upwards, instead of downwards, 'tis probable, that in that case the upper part of each ringlet of bark, and not the lower, would have swelled, by having nourishment thereby brought to it from the inmost wood.

We may hence also see the reason why, when a tree is unfruitful, it is brought to bear fruit, by the taking ringlets of bark off from its branches, *viz.* because thereby a less quantity of sap arising, it is better digested and prepared for the nourishment of the fruit; which from the greater quantity of oil, that is usually found in the seeds, and their containing vessels, than in other parts of plants, shows that more sulphur and air is requisite for their production, than there is for the production of wood and leaves.

But the most considerable objection against this progressive motion of the sap, without a circulation, arises from hence, *viz.* that it is too precipitate a course, for a due digestion of the sap, in order to nutrition: Whereas in animals nature has provided, that many parts of the blood shall run a long course, before they are either applied to nutrition, or discharged from the animal.

But when we consider, that the great work of nutrition, in vegetables as well as animals, (I mean after the nutriment is got into the veins and arteries of animals) is chiefly carried on in the fine capillary vessels, where nature selects and combines, as shall best suit her different purposes, the several mutually attracting nutritious particles which were hitherto kept disjoined by the motion of their fluid vehicle; we shall find that nature has made an abundant provision for this work in the structure of vegetables; all whose composition is made up of nothing else but innumerable fine capillary vessels, and glandulous portions or vesicles.

Upon the whole, I think we have, from these experiments and observations, sufficient ground to believe that there is no circulation of the sap in vegetables; notwithstanding many ingenious persons have been induced to think there was, from several curious observations and experiments, which evidently prove, that the sap does in some measure recede from the top towards the lower parts of plants, whence they were with good probability of reason induced to think that the sap circulated.

The likeliest method effectually and convincingly to determine this difficulty, whether the sap circulates or not, would be by ocular inspection, if that could be attained: And I see no reason we have to despair of it, since by the great quantities imbibed and perspired, we have good ground to think, that the progressive motion of the sap is considerable in the largest sap vessels of the transparent stems of leaves: And if our eyes, assisted with microscopes, could come at this desirable sight, I make no doubt but that we should see the sap, which was progressive in the heat of the day, would on the coming on of the cool evening, and the falling dew, be retrograde in the same vessels.

CHAP. V.

Experiments, whereby to prove, that a considerable quantity of air is inspired by Plants.

It is well known that air is a fine elastick fluid, with particles of very different natures floating in it, whereby it is admirable fitted by the great author of nature, to be the breath of life, of vegetables, as well as of animals, without which they can no more live, nor thrive than animals can.

In the Experiments on Vines, chapter III. we saw the very great quantity of air which was continually ascending from the Vines, thro' the sap in the tubes; which manifestly shews what plenty of it is taken in by vegetables, and is perspired off with the sap thro' the leaves.

EXPERIMENT XLVII.

Sept. 9th, at 9 *a. m.* I cemented an *Apple-branch b* (Fig. 11.) to the glass tube *r i e z*: I put no water in the tube, but set the end of it in the cistern of water *x*. Three hours after I found the water sucked up in the tube many inches to *z*; which shews, that a considerable quantity of air was imbibed by the branch, out of the tube *r i e z*: And in like manner did the Apricot-branch (Exper. 29.) daily imbibe air.

EXPERIMENT XLVIII.

I took a cylinder of Birch with the bark on, 16 inches long and $\frac{3}{4}$ diameter, and cemented it fast at *z;* (Fig. 32.) to the hole in the top of the air pump receiver *p p*, setting the lower end of it in the cistern of water *x*; the upper end of it at *n* was well closed up with melted cement.

I then drew the air out of the receiver, upon which innumerable air bubbles issued continually out of the stick into the water *x*. I kept the receiver exhausted all that day, and the following night, and till the next day at noon, the air all the while issuing into the water *x*: I continued it thus long in this state, that I might be well assured, that the air must pass in through the bark, to supply that great and long flux of air at *x*. I then cemented up 5 old eyes in the stick, between *z* and *n*, where little shoots had formerly been, but were now perished, yet the air still continued to flow freely at *x*.

It was observable in this, and many of the Experiments on sticks of other trees, that the air which could enter only thro' the bark between *z* and *n*, did not issue into the water, at the bottom of the stick, only at or near the bark, but thro' the whole and inmost substance of the wood, and that chiefly, as I guess by the largeness of the bases of the hemispheres of air thro' the largest vessels of the wood; which observation corroborates Dr. *Grew*'s and *Malpighi*'s opinion, that they are air vessels.

I then cemented upon the receiver the cylindrical glass *y y*, and filled it full of water, so as to stand an inch above the top *n* of the stick.

The air still continued to flow at *x*, but in an hour's time it very much abated, and in two hours ceased quite; there being now no passage of fresh air to enter, and supply what was drawn out of the stick.

I then, with a glass crane drew off the water out of the cylinder *y y*, yet the air did not issue thro' the wood at *x*.

I therefore took the receiver with the stick in it, and held it near the fire, till the bark was well dryed; after which I set it upon the air pump, and exhausted the air, upon which the air issued as freely at *x* as it did before the bark had been wetted, and continued so to do, tho' I kept the receiver exhausted for many hours.

I fixed in the same manner as the preceding Birch stick,

three joynts of a *Vine-branch*, which was two years old, the uppermost knot *r* being within the receiver; when I pumped the air passed most freely into the water *x x*.

I cemented fast the upper end of the stick *n* and then pumped, the air still issued out at *x*, tho' I pumped very long, but there did not now pass the 20th part of the air which passed when the end *n* was not cemented.

I then inverted the stick, placing *n* six inches deep in the water, and covered all the bark from the surface of the water to *z* the top of the receiver with cement; then pumping the air which entered at the top of the stick, passed thro' the immersed part of the bark: When I ceased pumping for some time, and the air had ceased issuing out; upon my repeating the pumping it would again issue out.

I found the same event in *Birch* and *Mulberry* sticks, in both which it issued most plentifully at old eyes, as if they were the chief breathing places for trees.

And Dr. *Grew* observes, that "the pores are so very large in the trunks of some plants, as in the better sort of thick walking canes, that they are visible to a good eye, without a glass; but with a glass the cane seems as if it were stuck top full of holes, with great pins, being so large as very well to resemble the pores of the skin, in the end of the fingers and ball of the hand.

"In the leaves of Pine they are likewise, through a glass, a very elegant shew, standing all most exactly in rank and file, thro' the length of the leaves." *Grew*'s Anatomy of Plants, *p.* 127.

Whence it is very probable, that the air freely enters plants, not only with the principal fund of nourishment by the roots, but also thro' the surface of their trunks and leaves, especially at night, when they are changed from a perspiring to a strongly imbibing state.

I fix'd in the same manner to the top of the air pump receiver, but without the cylindrical glass *y y*, the young shoots of the *Vine*, *Apple-tree*, and *Honysuckle*, both erected and inverted, but found little or no air came either from branches or leaves, except what air lay in the furrows, and the innumerable little pores of the leaves, which are plainly visible with the microscope. I tried also the single leaf of a *Vine*, both by immersing the leaf in the water *x*, and letting the stalk stand out of the receiver, as also by placing the leaf out of the receiver, and the stalk in the glass of water *x*; but little or no air came either way.

I observe in all these Experiments, that the air enters very slowly at the bark of young shoots and branches, but much more freely thro' old bark: And in different kinds of trees it has very different degrees of more or less free entrance.

I repeated the same Experiment upon several roots of trees: The air passed most freely from *n* to *x*; and when the glass vessel *y y* was full of water, and there was no water in *x*, the water passed at the rate of 3 ounces in 5 minutes; when the upper end *n* was cemented up, and no water in *y y*, some air, tho' not in great plenty, would enter the bark at *z f*, and pass thro' the water at *x*.

And that there is some air both in an elastick and unelastick state, mix'd with the earth, (which may well enter the roots with the nourishment) I found by putting into the inverted glass *z z a a* full of water (Fig. 35.) some earth dug up in an alley in the garden, which after it had stood soaking for several days, yielded a little elastick air, tho' the earth was not half dissolved. And in Experiment 68. we find that a cubick inch of earth yielded 43 cubick inches of air by distillation, a good part of which was roused by the action of the fire from a fix'd to an elastick state.

I fixed also in the same manner young tender fibrous roots, with the small end upwards at *n*, and the vessel *y y* full of water; then upon pumping large drops of water followed each other fast, and fell into the cistern *x*, which had no water in it.

CHAP. VI.

A specimen of an attempt to analyze the Air by a great variety of chymio-statical Experiments, which shew in how great a proportion Air is wrought into the composition of animal, vegetable, and mineral Substances, and withal how readily it resumes its former elastick state, when in the dissolution of those Substances it is disingaged from them.

Having in the preceding chapter produced many Experiments, to prove that the Air is freely inspired by Vegetables, not only at their roots, but also thro' several parts of their trunks and branches, which Air was most visibly seen ascending in great plenty thro' the sap of the Vine, in tubes which were affixed to them in the bleeding season; this put me upon making a more particular enquiry into the nature of a Fluid, which is so absolutely necessary for the support of the life and growth of Animals and Vegetables.

The excellent Mr. *Boyle* made many Experiments on the Air, and among other discoveries, found that a good quantity of Air was producible from Vegetables, by putting Grapes, Plums, Gooseberries, Cherries, Pease, and several other sorts of fruits and grains into exhausted and unexhausted receivers, where they continued for several days emitting great quantities of Air.

Being desirous to make some further researches into this matter, and to find what proportion of this Air I could obtain out of the different substances in which it was lodged and incorporated, I made the following chymio-statical Experiments: For, as whatever advance has here been made in the knowledge of the nature of Vegetables, has been owing to statical Experiments, so since nature, in all her operations, acts conformably to those mechanick laws, which were established at her first institution; it is therefore reasonable to conclude, that the likeliest way to enquire, by chymical operations, into the nature of a fluid, too fine to be the object of our sight, must be by finding out some means to estimate what influence the usual methods of analysing the animal, vegetable, and mineral kingdoms, has on that subtile fluid; and this I effected by affixing to retorts and boltheads hydrostatical gages, in the following manner, *viz.*

In order to make an estimate of the quantity of Air, which arose from any body by distillation or fusion, I first put the matter which I intended to distill into the small retort *r* (Fig. 33.) and then at *a* cemented fast to it the glass vessel *a b*, which was very capacious at *b*, with a hole in the bottom. I bound bladder over the cement which was made of tobacco-pipe clay and bean flower, well mixed with some hair, tying over all four small sticks, which served as splinters to strengthen the joynt; sometimes, instead of the glass vessel *a b*, I made use of a large bolthead, which had a round hole cut, with a red hot iron ring at the bottom of it; through which hole was put one leg of an inverted syphon, which reached up as far as *z*. Matters being thus prepared, holding the retort uppermost, I immersed the bolthead into a large vessel of water, to *a* the top of the bolthead; as the water rushed in at the bottom of the bolthead, the Air was driven out thro' the syphon: When the bolthead was full of water to *z*, then I closed the outward orifice of the syphon with the end of my finger, and at the same time drew the other leg of it out of the bolthead, by which means the water continued up to *z*, and could not subside. Then I placed under the bolthead, while it was in the water, the vessel *x x*, which done, I lifted the vessel *x x* with the bolthead in it out of the water, and tyed a waxed thread at *z* to mark the height of the water: And then approached the retort gradually to the fire, taking care to screen the whole bolthead from the heat of the fire.

The descent of the water in the bolthead shewed the sums of the expansion of the Air, and of the matter which was distilling: The expansion of the Air alone, when the lower part of the retort was beginning to be red hot, was at a medium, nearly equal to the capacity of the retorts, so that it then took up a double space; and in a white and almost melting heat, the Air took up a tripple space or something more: for which reason the least retorts are best for these Experiments. The expansion of the distilling bodies was sometimes very little, and sometimes many times greater than that of the Air in the retort, according to their different natures.

When the matter was sufficiently distilled, the retort *&c.* was gradually removed from the fire, and when cool enough, was carried into another room, where there was no fire. When all was thoroughly cold, either the following day, or sometimes 3 or 4 days after, I marked the surface of the water *y*, where it then stood; if the surface of the water was below *z*, then the empty space between *y* and *z* shewed how much Air was generated, or raised from a fix'd to an elastick state, by the action of the fire in distillation: But if *y* the surface of the water was above *z*, the space between *z* and *y*, which was filled with water, shewed the quantity of Air which had been absorbed in the operation, *i.e.* was changed from a repelling elastick to a fix'd state, by the strong attraction of other particles, which I therefore call absorbing.

When I would measure the quantity of this new generated Air, I separated the bolthead from the retort, and putting a cork into the small end of the bolthead, I inverted it, and poured in water to *z*. Then from another vessel (in which I had a known quantity of water by weight) I poured in water to *y*; so the quantity of water which was wanting, upon weighing this vessel again, was equal to the bulk of the new generated Air. I chose to measure the quantities of Air, and the matter from whence it arose, by one common measure of cubick inches, estimated from the specifick gravities of the several substances, that thereby the proportion of one to the other might the more readily be seen.

I made use of the following means to measure the great quantities of Air, which were either raised and generated, or absorbed by the fermentation arising from the mixture of variety of solid and fluid substances, whereby I could easily estimate the surprising effects of fermentation on the air, *viz.*

I put into the bolthead *b* (Fig. 34.) the ingredients, and then run the long neck of the bolthead into the deep cylindrical glass *a y*, and inclined the inverted glass *a y*, and bolthead almost horizontally in a large vessel of water, that the water might run into the glass *a y*; when it was almost up to *a* the top of the bolthead, I then immersed the bottom of the bolthead, and lower part *y* of the cylindrical glass under water, raising at the same time the end *a* uppermost. Then before I took them out of the water, I set the bolthead and lower part of the cylindrical glass *a y* into the earthen vessel *x x* full of water, and having lifted all out of the great vessel of water, I marked the surface *z* of the water in the glass *a y*.

If the ingredients in the bolthead, upon fermenting generated Air, then the water would fall from *z* to *y*, and the empty space *z y* was equal to the bulk of the quantity of Air generated: But if the ingredients upon fermentation did absorb or fix the active particles of Air, then the surface of the water would ascend from *z* to *n*, and the space *z n*, which was filled with water, was equal to the bulk of Air, which was absorbed by the ingredients, or by the fume arising from them: When the quantities of Air, either generated or absorbed, were very great, then I made use of large chymical receivers instead of the glass *a y*: But if these quantities were very small, then instead of the bolt-head and deep cylindrical glass *a y*, I made use of a small cylindrical glass, or a common beer glass inverted, and placed under it a Viol or Jelly glass, taking care that the water did not come at the ingredients in them, which was easily prevented by drawing the water up under the inverted glass to what height I pleased by means of a syphon; I measured the bulk of the spaces *z y* or *z n*, by pouring in a known quantity of water, as in the foregoing Experiment, and making an allowance for the bulk of the neck of the bolthead, within the space *z y*.

When I would take an estimate of the quantity of Air absorbed and fix'd, or generated by a burning candle, burning brimstone or nitre, or by the breath of a living animal, *&c.* I first placed a high stand, or pedestal in the vessel full of water *x x*; (Fig. 35.) which pedestal reached a little higher than *z z*. On this pedestal I placed the candle, or living animal, and then whelmed over it the large inverted glass *z z a a*, which was suspended by a cord, so as to have its mouth *r r* three or four inches under water; then with a syphon I sucked the Air out of the glass vessel till the water rose to *z z*. But when any noxious thing, as burning brimstone, aquafortis, or the like, were placed under the glass; then by affixing to the syphon the nose of a large pair of bellows, whose wide sucking orifice was closed up, as the bellows were enlarged, they drew the Air briskly out of the glass *z z a a* thro' the syphon; the other leg of which syphon I immediately drew from under the glass vessel, marking the height of the water *z z*.

When the materials on the pedestal generated Air, then the water would subside from *z z* to *a a*, which space *z z a a* was equal to the quantity of Air generated: But when the materials destroyed any part of the Air's elasticity, then the water would rise from *a a* (the height that I in that case at first sucked it to) to *z z*, and the space *a a z z* was equal to the quantity of Air, whose elasticity was destroyed.

I sometimes fired the materials on the pedestal by means of a burning glass, *viz.* such as phosphorus and brown paper dipped in water, strongly impregnated with nitre and then dryed.

Sometimes I lighted the candle or large matches of brimstone before I whelmed the glass *z z a a* over them; in which case I instantly drew up the water to *a a*, which by the expansion of the heated Air would at first subside a little, but then immediately turned to rising state, notwithstanding the flame continued to heat and rarify the Air for 2 or 3 minutes: As soon as the flame was out, I marked the height of the water *z z*; after which the water would for 20 or 30 hours continue rising a great deal above *z z*.

Sometimes when I would pour violently fermenting liquors, as aquafortis, *&c.* on any materials, I suspended the aquafortis in a viol at the top of the glass vessel *z z a a*, in such manner, that by means of a string, which came down into the vessel *x x*, I could by inverting the viol pour the aquafortis on the materials, which were in a vessel on the pedestal.

I shall now proceed to give an account of the event of a great many Experiments, which I made by means of these instruments, which I have here at first described, to avoid the frequent repetition of a description of 'em. It is consonant to the right method of philosophising, first, to analise the subject, whose nature and properties we intend to make any researches into, by a regular and numerous series of Experiments: And then by laying the event of those Experiments before us in one view, thereby to see what light their united and concurring evidence will give us. How rational this method is, the sequel of these Experiments will shew.

The illustrious Sir *Isaac Newton* (query 31st of his Opticks) observes, that "true permanent Air arises by fermentation or heat, from those bodies which the chymists call fixed, whose particles adhere by a strong attraction, and are not therefore separated and rarified without fermentation. Those particles receding from one another with the greatest repulsive force, and being most difficultly brought together, which upon contact were most strongly united. And query 30. dense bodies by fermentation rarify into several sorts of Air; and this Air by fermentation, and sometimes without it, returns into dense bodies." Of the truth of which we have evident proof from many of the following Experiments, *viz.*

That I might be well assured that no part of the new Air which was produced in distillation of bodies, arose either from the greatly heated Air in the retorts, or from the substance of the heated retorts, I first gave a red hot heat both to an empty glass retort and also to an iron retort made of a musket barrel; when all was cold, I found the Air took up no more room than before it was heated: whence I was assured, that no Air arose, either from the substance of the retorts, or from the heated Air.

As to animal substances, a very considerable quantity of permanent Air was produced by distillation, not only from the blood and fat, but also from the most solid parts of animals.

Experiment XLIX.

A cubick inch of *Hog's blood*, distilled to dry scoria, produced thirty three cubick inches of Air, which Air did not arise till the white fumes arose; which was plain to be seen by the great descent of the water at that time, in the receiver *a z y* (Fig. 33.)

Experiment L.

Less than a cubick inch of *Tallow*, being all distilled over into the receiver *a z y* (Fig. 33.) produced 18 cubick inches of Air.

Experiment LI.

241 Grains, or half a cubick inch of the tip of a *fallow Deer's horn*, being distilled in the iron retort, made of a musket barrel, which was heated at a smith's forge, produced 117 cubick inches, that is, 234 times its bulk of Air, which did not begin to rise till the white fumes arose; but then rushed forth in great abundance, and in good plenty, also with the fœtid oil which came last. The remaining calx was two thirds black, the rest ash-coloured; it weighed 128 grains, so it was not half wasted, whence there must remain much sulphur in it; the weight of water to Air, being nearly as 885 to one, as Mr. *Hawksbee* found it, by an accurate Experiment. A cubick inch of Air will weigh $\frac{2}{7}$ of a grain, whence the weight of Air in the horn was 33 grains, that is, near $\frac{1}{7}$ part of the whole horn.

We may observe in this, as also in the preceding Experiment, and many of the following ones, that the particles of new Air were detached from the blood and horn, at the same time with the white fumes, which constitute the volatile salt: But this volatile salt, which mounts with great activity in the Air, is so far from generating true elastick Air, that on the contrary it absorbs it, as I found by the following Experiment.

Experiment LII.

A dram of *volatile salt of sal armoniack*, soon distilled over with a gentle heat; but tho' the expansion in the receiver was double that of heated Air alone, yet no Air was generated, but two and an half cubick inches were absorbed.

Experiment LIII.

Half a cubick inch of *Oystershell*, or 266 grains distilled in the iron retort, generated 162 cubick inches, or 46 grains, which is a little more than $\frac{1}{6}$ part of the weight of the shell.

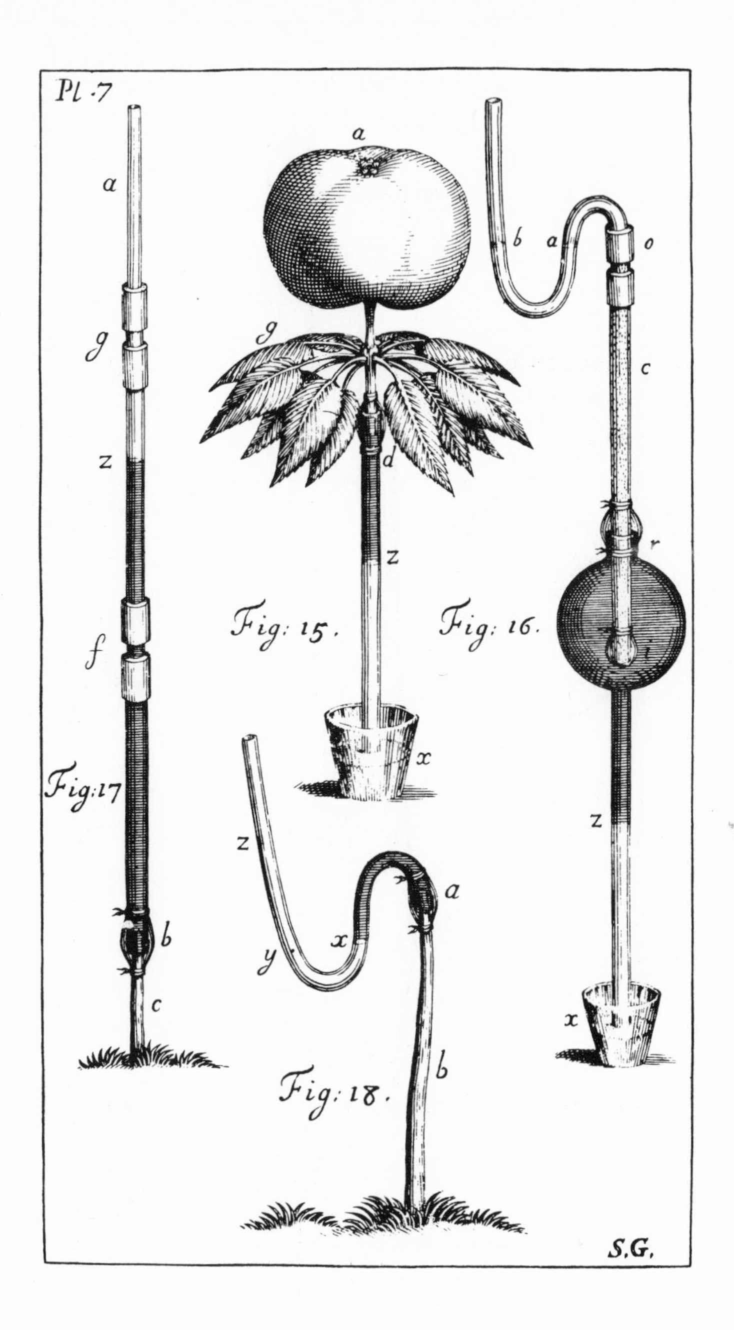
Pl. 7
a
g
z
f
Fig: 17
b
c
a
g
d
z
Fig: 15.
x
z
x
a
y
b
Fig: 18.
b
a
o
c
r
Fig: 16.
i
z
x
S.G.

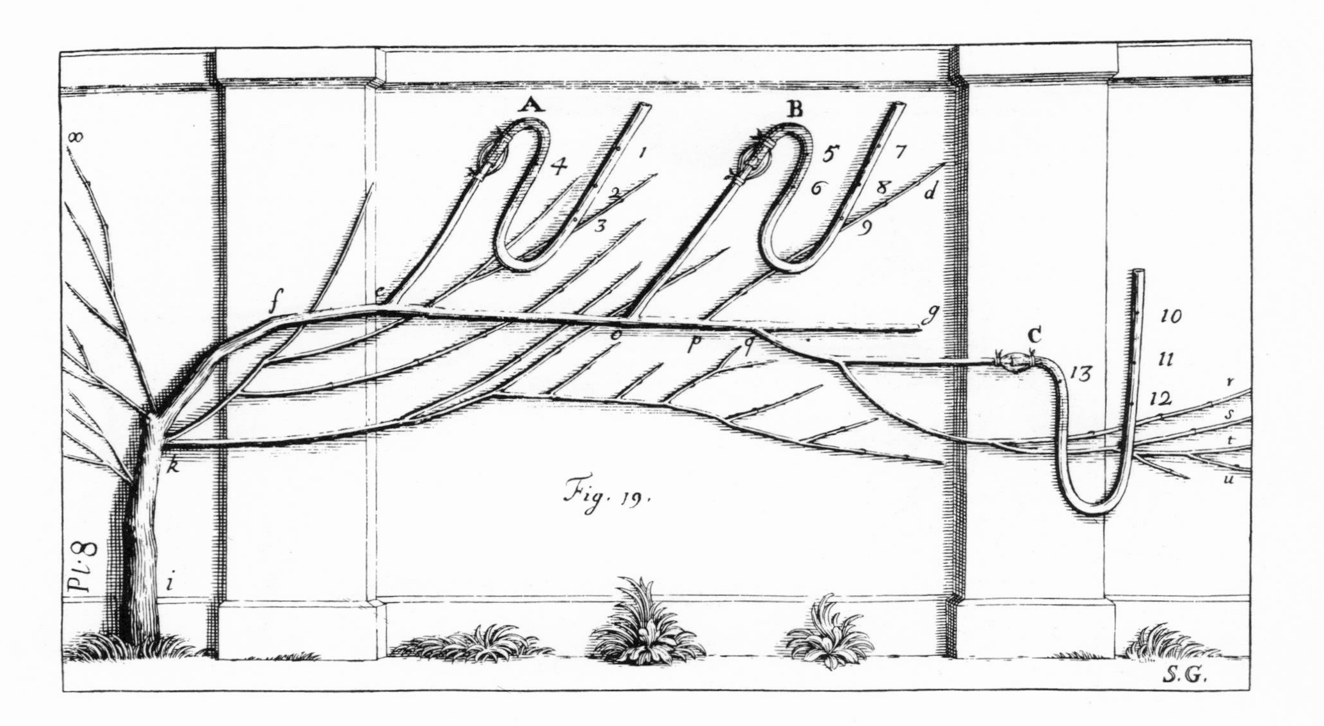

Fig. 19.

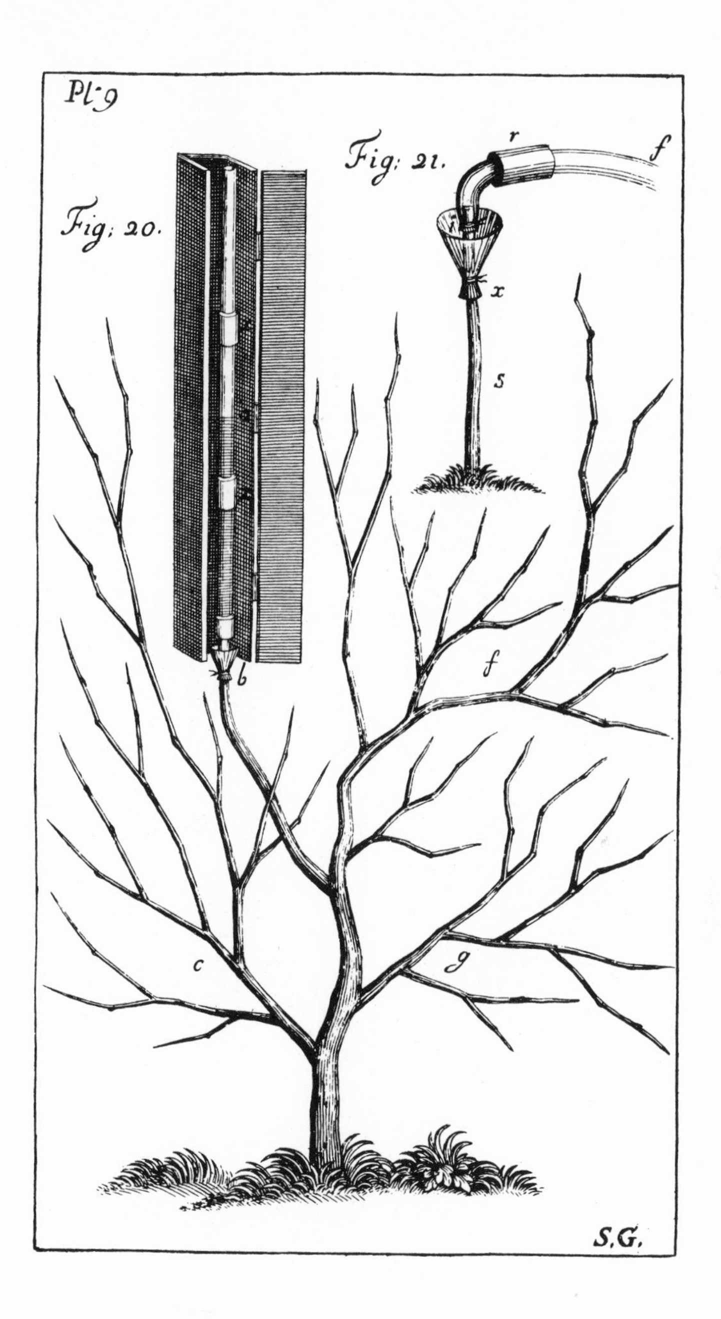
Pl. 9
Fig: 20.
Fig: 21.
r
f
i
x
s
b
f
c
g
S.G.

Pl·10
Fig: 23.
a
b
c
x
x
Fig: 22.
f
e
h
g
b
S.G.

Experiment LIV.

Two grains of *Phosphorus* easily melted at some distance from the fire, flamed and filled the retort with white fumes, it absorbed three cubick inches of Air. A like quantity of *Phosphorus*, fired in a large receiver (Fig. 35.) expanded into a space equal to sixty cubick inches, and absorbed 28 cubick inches of Air: When 3 grains of *Phosphorus* were weighed, soon after it was burnt, it had lost half a grain of its weight; but when two grains of *Phosphorus* were weighed, some hours after it was burnt, having run more *per deliquium* by absorbing the moisture of the Air, it had increased a grain in weight.

Experiment LV.

As to vegetable Substances, from half a cubick inch, or 135 grains of heart of *Oak*, fresh cut from the growing tree, was generated 108 cubick inches of Air, *i. e.* a quantity equal to 216 times the bulk of the piece of *Oak*, its weight was above 30 grains, $\frac{1}{4}$ part of the weight of 135 grains of *Oak*. I took a like quantity of thin shavings from the same piece of *Oak*, and dryed them gently at some distance from a fire for 24 hours, in which time 44 grains weight of moisture had evaporated; which being deducted from the 135 grains, there remains 91 grains for the solid part of the *Oak*: Then the 30 grains of Air, will be $\frac{1}{3}$ of the weight of the solid part of the *Oak*.

Eleven days after this Air was made, I put a live Sparrow into it, which died instantly.

Experiment LVI.

From 388 grains weight of *Indian Wheat*, which grew in my garden, but was not come to full maturity, was generated 270 cubick inches of air, the weight of which air was 77 grains, *viz.* $\frac{1}{4}$ of the weight of the Wheat.

EXPERIMENT LVII.

From a cubick inch, or 318 grains of *Pease*, was generated 396 cubick inches of Air or 113 grains, *i. e.* something more than $\frac{1}{3}$ of the weight of the *Pease*.

Nine days after this Air was made, I lifted the inverted mouth of the receiver which contained it, out of the water, and put a lighted candle under it, upon which it instantly flashed: Then I immediately immersed the mouth of the receiver in the water, to extinguish the flame. This I repeated 8 or 10 times, and it as often flashed, after which it ceased, all the sulphureous spirit being burnt. It was the same with air of distilled Oystershell and Amber, and with new distilled air of Pease and Bees-wax. I found it the same also with another like quantity of Air of Pease; notwithstanding I washed that Air no less than eleven times, by pouring it so often under water, upwards, out of the containing vessel, into another inverted receiver full of water.

EXPERIMENT LVIII.

There was raised from an ounce or 437 grains, of *Mustard-seed* 270 cubick inches of air, or 77 grains, which is something more than $\frac{1}{6}$ part of the ounce weight. There was doubtless much more air in the feed; but it rose in an unelastick state, being not disentangled from the Oil, which was in such plenty within the gun-barrel, that when I heated the whole barrel red hot in order to burn it out, it flamed vigorously out at the mouth of the barrel. Oil also adhered to the inside of the barrel, in the distillation of many of the other animal, vegetable and mineral substances; so that the elastick air, which I measured in the receiver, was not all the air contained in the several distill'd substances; some remaining in the Oil, for there is unelastick air in Oil, part being also resorbed by the sulphureous fumes in the receiver.

Experiment LIX.

From half a cubick inch of *Amber*, or 135 grains, was raised 135 cubick inches of air, or 38 grains, *viz.* $\frac{1}{3.55}$ part of its weight.

Experiment LX.

From 142 grains of dry *Tobacco* was raised 153 cubick inches of air, which is little less than $\frac{1}{3}$ of the whole weight of the Tobacco; yet it was not all burnt, part being out of the reach of the fire.

Experiment LXI.

Camphire is a most volatile sulphureous substance sublimed from the Rosin of a tree in the *East-Indies*. A dram of it melted into a clear liquor, at some distance from the fire, and sublimed in the form of white chrystals, a little above the liquor, it made a very small expansion, and neither generated nor absorbed air. The same Mr. *Boyle* found, when he burnt it *in vacuo,* Vol. 2, *p.* 605.

Experiment LXII.

From about a cubick inch of chymical *Oil of Aniseed*, I obtained 22 cubick inches of air; and from a like quantity of Oil of Olives 88 cubick inches of air. The reason of which difference was, as I suppose, this, *viz.* finding that the Oil of Anniseed came plentifully over into the receiver, in the distillation of the Oil of Olives, I raised the neck of the retort a foot higher, by which means the Oil could not so easily ascend, but fell back again into the hotest part of the retort, whereby more air was separated; yet in this case good store of Oil came over into the receiver; in which there was doubtless plenty of unelastick air: Whence by comparing this with Experiment 58. we see that air is in greater plenty separated from the Oil, when in the Mustard-seed, than it is from expressed or chymical Oil.

EXPERIMENT LXIII.

From a cubick inch, or 359 grains of *Honey*, mixed with calx of bones, there arose 144 cubick inches of air, or 41 grains, *viz.* a little more than $\frac{1}{9}$ part of the weight of the whole.

EXPERIMENT LXIV.

From a cubick inch of yellow *Bees-wax*, or 243 grains, there arose 54 cubick inches of air, or 15 grains; the $\frac{1}{16}$ part of the whole.

EXPERIMENT LXV.

From 373 grains, or a cubick inch of the coarsest *Sugar*, which is the essential salt of the sugar-cane, there arose 126 cubick inches of air, equal to 36 grains, a little more than $\frac{1}{10}$ part of the whole.

EXPERIMENT LXVI.

I found very little air in 54 cubick inches of *Brandy*, but in a like quantity of *Well-water* I found one cubick inch. In *Piermont-water* there is near twice as much air, as in *Rain* or *common water*, which air contributes to the briskness of that and many other mineral waters. I found these several quantities of air, in these waters, by inverting the noses of bottles, full of these several liquors, into small glass cisterns full of the same liquor. And then setting them all together in a boyler, where having an equal heat, the air was thereby separated and ascended to the upper parts of the bottles.

EXPERIMENT LXVII.

By the same means also, I found plenty of air might be obtained from *minerals*. Half a cubick inch, or 158 grains of *Newcastle coal*, yielded 180 cubick inches of air, which arose very fast from the coal, especially when the yellowish fumes ascended. The weight of this air is 51 grains, which is nearly $\frac{1}{3}$ of the weight of the coals.

EXPERIMENT LXVIII.

A cubick inch of fresh dug *untried Earth* off the common, being well burnt in distillation, produced 43 cubick inches of air. From *chalk* also, I obtained air in the same manner.

EXPERIMENT LXIX.

From a quarter of a cubick inch of *Antimony*, I obtain'd 28 times its bulk of air. It was distilled in a glass retort, because it will demettalize iron.

EXPERIMENT LXX.

I procured a hard, dark, gray *Pyrites, a mineral substance*, which was found 7 feet under ground, in digging for springs on *Walton-Heath*, for the service of the Right Honourable the Earl of *Lincoln*, at his beautiful Seat at *Oatlands* in *Surrey*; this mineral abounds not only with sulphur, which has been drawn from it in good plenty, but also with saline particles, which shoot visibly on its surface. A cubick inch of this *mineral* yielded in distillation 83 cubick inches of air.

EXPERIMENT LXXI.

Half a cubick inch of well decrepitated *sea-salt*, mixt with double its quantity of calx of bones generated 32 times

its bulk of air: It had so great a heat given it, that all being distilled over, the remaining scoria did not run *per deliquium*. I cleared the gun-barrel of these and the like scoria, by striking long on the outside with a hammer.

Experiment LXXII.

From 211 grains or half a cubick inch of *Nitre*, mixed with calx of bones, there arose 90 cubick inches of air, *i. e.* a quantity equal to 180 times its bulk; so the weight of air in any quantity of nitre is about $\frac{1}{8}$ part. *Vitriol* distilled in the same manner yields air too.

Experiment LXXIII.

From a cubick inch or 443 grains of *Renish Tartar*, there arose very fast 504 cubick inches of air; so the weight of the air in this Tartar was 144 grains, *i. e.* $\frac{1}{3}$ part of the weight of the whole: The remaining scoria, which was very little, run *per deliquium*, an argument that there remained some *Sal Tartar*, and consequently more air; for

Experiment LXXIV.

Half a cubick inch or 304 grains of *Sal Tartar*, made with nitre and tartar, and mixed with a double quantity of calx of bones, yielded in distillation 112 cubick inches of air; that is, 224 times its bulk of air, which 112 cubick inches weighing 32 grains, is nearly $\frac{1}{9}$ part of the weight of the *Sal Tartar*. There is a more intense degree of heat required to raise the air from *Sal Tartar* than from nitre.

Hence we see, that the proportion of air in equal bulks of *Sal Tartar* and nitre is as 224 to 180. But weight for weight, nitre contains a little more air in it, than this *Sal Tartar* made with nitre. But *Sal Tartar* made without nitre, has probably a little more air in it than this had, because it is found to make a greater explosion in the *Pulvis Fulminans*, than the nitrated *Sal Tartar*. But supposing, as is found by this Experiment, that *Sal Tartar*, according to its specifick gravity, contains $\frac{1}{5}$ part more in it than nitre; yet this excess of air is not sufficient to account for the vastly greater explosion of *Sal Tartar* than of nitre; which seems principally to arise from the more fixt nature of *Sal Tartar*; which therefore requires a more intense degree of fire, to separate the air from the strongly adhering particles, than is found requisite to raise the air from nitre. Whence the air of *Sal Tartar* must necessarily thereby acquire a greater elastick force, and make a more violent explosion, than that of nitre. And from the same reason it is, that *Aurum Fulminans* gives a louder explosion than *Pulvis Fulminans*. The scoria of this operation did not run *per deliquium*, a proof that all the *Sal Tartar* was distilled over.

From the little quantity of air which is obtained by the distillation of sea-salt in Experiment 71. in comparison of what arises from nitre and *Sal Tartar*, we see the reason why it will not go off with an explosive force, like those when fired. And at the same time we may hence observe, that the air included in nitre and *Sal Tartar*, bears a considerable part in their explosion. For sea-salt contains an acid spirit as well as nitre; and yet that without a greater proportion of air does not qualify it for explosion, tho' mixed like nitre in the composition of gun-powder, with sulphur and charcoal.

Mr. *Boyle* found that *Aqua-fortis* poured on a strong solution of salt of tartar did not shoot into fair crystals of salt-petre, till it had been long exposed to the open air; whence he suspected that the air contribution to that artificial production of salt-petre. And says, "whatever the air hath to do in this Experiment, we have known such changes made in some saline concretes, chiefly by the help of the open air, as very few would be apt to imagine." Vol 1. p. 302. and Vol 3. p. 80.

We see from the great quantity of air, which is found in salts, of what use it is in their crystallization and formation, and particularly how necessary it is in making salt-petre from the mixture of salt of tartar and spirit of nitre. For since by Experiment 72 and 73, a great deal of air flies away, in the making of *Sal Tartar*, either from nitre and tartar, or from tartar alone: It must needs be necessary, in order to the forming of nitre from the mixture of *Sal Tartar* and spirit of nitre, that more air should be incorporated with it, than is contained either in the *Sal Tartar* or spirit of nitre.

Experiment LXXV.

Near half a cubick inch of *compound Aqua-fortis*, which bubbled and made a considerable expansion in distillation was soon distilled off: As it cooled the expansion abated very fast, and a little air was absorbed. Whence it is evident that the air generated by the distillation of nitre, did not arise from the volatile spirituous particles.

Hence also it is probable that there is some air in acid spirits, which is resorbed and fixt by them in distillation. And this is further confirmed from the many air bubbles which arise from *Aqua-regia*, in the solution of gold; for since gold loses nothing of its weight in being dissolved, the air cannot arise from the metalline part of the gold, but must either arise from the *Aqua-regia* or from latent air in the pores of the gold.

Experiment LXXVI.

A cubick inch of common *Brimstone* expanded very little in distillation in a glass retort; notwithstanding it had a great heat given it, and was all distilled over into the receiver without flaming. It absorbed some air, but flaming brimstone by Experiment 103, absorbs much air.

A good part of the air thus raised from several bodies by the force of fire, was apt gradually to lose its elasticity, in standing several days; the reason of which was (as will appear more fully hereafter) that the acid sulphureous fumes raised with that air, did resorb and fix the elastick particles.

Experiment LXXVII.

To prevent which I made use of the following method of distillation, *viz.* I fixt a leaden syphon, Fig. 38. to the nose of the iron retort *r r*; and then having immersed the syphon in the vessel of water *x x*, I placed over the open end of the syphon the inverted chymical receiver *a b* which was full of water; so that as the air which was raised in distillation, passed thro' the water up to the top of the receiver *a b*, a good part of the acid spirit and sulphureous fumes were by this means intercepted and retained in the water; the consequence of which was, that the new generated air continued in a more permanently elastick state, very little of it losing its elasticity, *viz.* not above a 15th or 18th part, and that chiefly the first 24 hours; after which the remainder continued in a constantly elastick state; excepting the air of tartar, which in 6 or 8 days lost constantly above one third of their elasticity; after which the remainder was permanently elastical.

That the great quantities of air which are thus obtained from these several substances by distillation are true air, and not a mere flatulent vapour, I was assured by the following tryals; *viz.* I filled a large receiver which contained 540 cubick inches, with air of tartar; and when it was cool, I suspended the receiver while its mouth was inverted in water. Then upon lifting the mouth of the receiver out of water, I immediately covered it by tying a piece of bladder over it. When I had found the exact weight, I blew out all the air of tartar with a pair of bellows which had a long additional nose that reached to the bottom of the receiver. And then tying the bladder on, I weighed it again, but could find no difference in the specifick gravity of the two airs, and it was the same with an air of tartar which was 10 days old.

As to the other property of the air, elasticity, I found it exactly the same in the air of tartar, which was 15 days old, and common air; by filling two equal tubes with these different airs, the tubes were 10 inches long and sealed at one end; I placed them at the same time in a cylindrical glass condensing receiver, where I compressed them with two additional atmospheres, taking care to secure myself from danger in case the glass should burst, by placing it in a deep wooden vessel, the water rose to equal heights in both tubes. This receiver was gently annealed and thereby toughened, by being boiled in Urine where it lay till all was cold.

I put also into the same tubes some new made air of tartar, both the tubes standing in cisterns of water; the air of one of these tubes I compressed in the condensing engine for some days, to try whether in that compressed state, more of the air's elasticity would be destroyed by the absorbing vapours than in an uncompressed state; but I did not perceive any sensible difference.

Lemery, in his course of chymistry, *p.* 592, obtained in the distillation of 48 ounces of *Tartar*, 4 ounces of phlegm, 8 of spirits, 3 of oil, and 32 of Scoria, *i. e.* two thirds of the whole, so one ounce was lost in the operation.

In my distillation of 443 grains of *Tartar* in Exper. 73. there remained but 42 grains of Scoria, which is little more than $\frac{1}{10}$ of the *Tartar*; and in this remainder there was by Exper. 74 air, for there was *Sal Tartar*, it running *per deliquium*.

Whence by comparing *Lemery's* and my distillation together, we shall find, that there remained in this 32 ounces of Scoria, and in the ounce that was lost, (which was doubtless most of it air) substance enough to account for the great quantity of air, which in Exper. 73, was raised from *Tartar*; especially, if we take into the account the proportion of air, which was contained in the oil which was $\frac{1}{16}$ part of the whole *Tartar*, for there is much air in oil.

The bodies which I distilled in this manner (Fig. 38.) were Horn, *calculus humanus*, Oystershell, Oak, Mustard seed, Indian-wheat, Pease, Tobacco, oil of Anniseed, oil of Olives, Honey, Wax, Sugar, Amber, Coal, Earth, *Walton* Mineral, sea Salt, Salt-petre, Tartar, *Sal Tartar*, Lead, Minium. The greatest part of the Air obtained from all which bodies was very permanent, except what the Air of Tartar lost in standing several days. Particularly, that from nitre lost little of its elasticity, whereas most of the Air obtained from nitre, in distilling with the receiver (Fig. 33.) was resorbed in a few days, as was also the Air which was generated from detonized nitre in Experiment 102. Hence also we see the reason, why 19 parts in 20, of the air which was generated, by the firing of Gunpowder, was in 18 days resorbed by the sulphureous fumes of the Gunpowder. As Mr. *Hawksbee* observed, in his physico-mechanical Experiments, *page* 83.

In the distillation of Horn, it was observed, that when towards the end of the operation the thick fœtid oil arose, it formed very large bubbles, with tough unctuous skins, which continued in that state some time; and when they broke, there arose out of them volumes of smoak, as out of a chimney, and it was the same in the distillation of Mustard-seed.

An account of some Experiments made on Stones taken out of human Urine and Gall Bladders.

Having, while these sheets were printing off, procured by the favour of Mr. *Ranby, Surgeon to His Majesty's Household*, some *calculi humani*, I made the following Experiments with them, which I shall here insert, *viz.*

I distilled a *calculus* in the iron retort (Fig. 38.) It weighed

230 grains, which is something less in bulk than $\frac{3}{4}$ of a cubick inch: There arose from it very briskly, in distillation, 516 cubick inches of elastick air, that is, a bulk equal to 645 times the bulk of the Stone; so that above half the Stone was raised by the action of the fire into elastick Air; which is a much greater proportion of Air, than I have ever obtained by fire, from any other substances, whether animal, vegetable or mineral. The remaining calx weighed 49 grains, that is, $\frac{1}{4{\cdot}69}$ part of the *calculus*; which is nearly the same proportion of calx, that the worthy Dr. *Slare* found remaining, after the distilling and calcining two ounces of *calculus*, "one ounce and three drams of which (he says) evaporated in the open fire (a material circumstance, which the Chymists rarely enquire after) of which we have no account." *Philos. Transact. Lowthorp's Abridgment.* Vol. III. p. 179. The greatest part of which was, we see by the present Experiment, raised into permanently elastick Air.

By comparing this distillation of the *calculus* with that of *Renish Tartar* in Exper. 73. we see that they both afford more Air in distillation, than any other substances: And it is remarkable, that a greater proportion of this new raised Air from these two substances, is resorbed, and loses its elasticity, in standing a few days, than that of any other bodies; which are strong symptoms that the *calculus* is a true *animal Tartar*. And as there was very considerably less oil, in the distillation of *Renish Tartar*, than there was in the distillation of the Seeds and solid parts of vegetables; so I found that this *calculus* contained much less oil than the blood or solid parts of animals.

I distilled in the same manner, as the above mentioned *calculus*, some stones taken out of a human gall bladder, they weighed 52 grains, so their bulk was equal to $\frac{1}{6}$ part of a cubick inch, as I found by taking their specifick gravity. There was 108 cubick inches of elastick Air raised from them in distillation, a quantity equal to 648 times their bulk; much the same quantity that was raised from the *calculus*. About

$\frac{1}{6}$ part of this elastick air was in 4 days reduced into a fix'd state. There arose much more oil in the distillation of these Stones, than from the *Calculus*, part of which oil did arise from the Gall which adhered to, and was dried on the surfaces of the Stones, which oil formed large bubbles, like those which arose in the distillation of Deer's Horn *p.* 187.

A small Stone of the Gall Bladder, which was as big as a Pea, was dissolved in a Lixivium of *Sal Tartar* in seven days, which Lixivium will also dissolve *Tartar*; yet it will not dissolve the *Calculus*, which is more firmly united in its parts.

A quantity of *Calculus* equal to one half of what was distilled, *viz.* 115 grains, did, when a cubick inch of spirit of nitre was poured on it, dissolve in 2 or 3 hours, with a large froth, and generated 48 cubick inches of Air, none of which lost its elasticity, tho' it stood many days in the glass vessel. (Fig. 34.) And a like quantity of *Tartar* being mixed with spirit of nitre, was in the same time dissolved, but no elastick Air was generated, notwithstanding *Tartar* abounds so much with air.

Small pieces of *Tartar* and *Calculus* were in 12 or 14 days both dissolved by oil of Vitriol; the like pieces of *Tartar* and *Calculus* were dissolved in a few hours by oil of Vitriol, into which there was gradually poured near an equal quantity of spirit of Harts-horn, made with Lime, which caused a considerable ebullition and heat.

Tho' the remaining calx of the distillation of *Tartar*, in Exper. 73. run *per deliquium*, and had therefore *Sal Tartar* in it; and tho' the calx of the distilled *Calculus* did not run *per deliquium*, and had consequently no *Sal Tartar* in it; yet it cannot thence be inferred, that the *Calculus* is not a tartarine substance: Because by Exper. 74. it is evident, that *Sal Tartar* it self, when mixed with an animal calx, distills all over, so that the calx will not afterwards run *per deliquium*.

By the great similitude there is therefore in so many

respects between these two substances, we may well look upon the *Calculus*, and the *Stone* in the *Gall Bladder*, as true *animal Tartars*, and doubtless *Gouty* concretions are the same.

From the great quantities of air that are found in these Tartars, we see that unelastick Air particles, which by their strongly attracting property are so instrumental in forming the nutritive matter of Animals and Vegetables, is by the same attractive power apt sometimes to form anomalous concretions, as the Stone, *&c.* in Animals, especially in those places where any animal fluids are in a stagnant state, as in the Urine and Gall Bladders. The like tartarine concretions are also frequently formed in some fruits, particularly in Pears; but they do then especially coalesce in greatest plenty, when the vegetable juices are in a stagnant state, as in wine vessels. *&c.*

This great quantity of strongly attracting, unelastick air particles, which we find in the *Calculus*, should rather encourage than discourage us, in searching after some proper dissolvent of the Stone in the Bladder, which, upon the analysis of it, is found to be well stored with active principles, such as are the principal agents in fermentation. For Mr. *Boyle* found therein a good quantity of volatile salt, with some oil, and we see by the present Experiment, that there is store of unelastick air particles in it. The difficulty seems chiefly to lay in the over proportion of these last mentioned particles, which are firmly united together by sulphur and salt, the proportion of *caput mortuum*, or earth being very small.

Experiment LXXVIII.

One eighth of a cubick inch of *Mercury* made a very insensible expansion in distillation, notwithstanding the iron retort had an almost melting heat given it, at a smith's forge, so that it made an ebullition, which could be heard at some distance, and withal shook the retort and receiver. There was no air generated, nor was there any expansion of air in the following Exper. *viz.*

Experiment LXXIX.

I put into the same retort half a cubick inch of *Mercury*, affixing to the retort a very capacious receiver, which had no hole in the bottom. The wide mouth of the receiver was adapted to the small neck of the retort (which was made of a musket barrel) by means of two large pieces of cork which entered and filled the mouth of the receiver, they having holes bored in them of a fit size for the neck of the retort; and the juncture was farther secured, by a dry supple bladder tyed over it: For I purposely avoided making use of any moist lute, and took care to wipe the inside of the receiver very dry with a warm cloth.

The *Mercury* made a great ebullition, and came some of it over into the receiver, as soon as the retort had a red heat given it, which was increased to a white and almost melting heat, in which state it continued for half an hour. During which time, I frequently cohobated some part of the *Mercury*, which condensed, and was lodged on an horizontal level, about the middle of the neck of the retort: And which upon raising the receiver, flowed down into the bottom of the retort, and there made a fresh ebullition, which had ceased, when all the *Mercury* was distilled from the bottom of the retort. When all was cool, I found about two drams of *Mercury* in the retort, and lost in the whole 43 grains, but there was not the least moisture in the receiver.

Whence it is to be suspected that Mr. *Boyle* and others were deceived by some unheeded circumstance, when they thought they obtained a water from *Mercury* in the distillation of it; which he says he did once, but could not make the like Experiment afterwards succeed. *Boyle* Vol. III. *p.* 416.

I remember that about 20 years since I was concerned with several others in making this Experiment at the elaboratory in Trinity College *Cambridge*, when imagining there would be a very great expansion, we luted a German earthen retort, to 3 or 4 large Alodals, and a capacious receiver; as Mr. *Wilson* did in his course of Chymistry. Four pounds of *Mercury* was poured by little and little into the red hot retort, thro' a tobacco-pipe purposely affixed to it. The event was, that we found some spoons full of water with the *Mercury* in the Alodals, which I then suspected to arise from the moisture of the earthen retort and lute, and am now confirmed in that suspicion. It rained incessantly all the day, when I made this present Experiment; so that when water is obtained in the distillation of *Mercury*, it cannot be owing to a moister temperature of the Air.

The effects of Fermentation on the Air.

Having from the foregoing Experiments seen very evident proof of the production of considerable quantities of true elastick air, from liquors and solid bodies, by means of fire; we shall find in the following Experiments many instances of the production; and also of the fixing or absorbing of great quantities of air by the fermentation arising from the mixture of variety of solids and fluids: Which method of producing and of absorbing, and fixing the elastick particles of air by fermentation, seems to be more according to nature's usual way of proceeding, than the other of fire.

EXPERIMENT LXXX.

I put into the bolthead *b* (Fig. 34.) 16 cubick inches of *Sheeps blood*, with a little water to make it ferment the better. I found by the descent of the water from *z* to *y* that in 18 days fourteen cubick inches of air were generated.

Experiment LXXXI.

Volatile Salt of Sal Ammoniac, placed in an open glass cistern, under the inverted glass *z z a a* (Fig. 35.) neither generated nor absorbed air. Neither did several other volatile liquors, as spirits of Harts-horn, spirits of Wine, nor compound Aquafortis, generate any air. But *Sal Ammoniac, Sal Tartar*, and spirits of Wine mixed together, generated 26 cubick inches of air, two of which was in 4 days resorbed, and after that generated again.

Experiment LXXXII.

Half a cubick inch of *Sal Ammoniac*, and double that quantity of *oil* of *Vitriol*, generated the first day 5 or 6 cubick inches: But the following days it absorbed 15 cubick inches, and continued many days in that state.

Equal quantities of spirits of *Turpentine*, and *oil of Vitriol*, had near the same effect, except that it was sooner in an absorbing state than the other.

Mr. *Geoffroy* shews, that the mixture of any vitriolic salts, with inflammable substances, will yield common Brimstone; and by the different compositions he has made of sulphur; and particularly from *oil* of *Vitriol*, and *oil* of *Turpentine*; and by the Analysis thereof, when thus prepared, he discovered it to be nothing but vitriolic salt, united with the combustible substance. *French* Memoirs, Anno 1704. *p.* 381, or *Boyle's* Works, Vol. III. *p.* 273. Notes.

Experiment LXXXIII.

In *February* I poured on six cubick inches of powdered *Oystershell*, an equal quantity of common white-wine

Vinegar. In 5 or 6 minutes it generated 17 cubick inches of air, and in some hours 12 cubick inches more, in all 29 inches. In nine days it had slowly resorbed 21 cubick inches of air. The ninth day I poured warm water into the vessel *x x*, (Fig. 34.) and the following day, when all was cool, I found that it had resorbed the remaining 8 cubick inches. Hence we see that warmth will sometimes promote a resorbing as well as a generating state, *viz.* by raising the resorbing fumes, as will appear more hereafter.

Half a cubick inch of *Oystershell*, and a cubick inch of *oil* of *Vitriol*, generated 32 cubick inches of air.

Oystershell, and 2 cubick inches of *sour Rennet*, of a Calve's stomach, generated in 4 days 11 cubick inches. But *Oystershell*, with some of the *liquor of a Calve's* stomach, which had fed much upon hay, did not generate air. It was the same with *Oystershell* and *Ox-gall*, *Urine* and *Spittle*.

Half a cubick inch of *Oystershell* and *Sevil* Orange juice generated the first day 13 cubick inches of air, and the following days it resorbed that, and 3 or 4 more cubick inches of air, and would sometimes generate again. It was the same with Limon juice.

Oystershell and *Milk* generated a little air: But *Limon juice* and *Milk* did at the same time absorb a little air; as did also *Calves Rennet* and *Vinegar*; some of the same *Rennet* alone generated a little air, and resorbed it again the following day. It had the same effect when mixed with *crums of bread*.

EXPERIMENT LXXXIV.

A cubick inch of *Limon juice*, and near an equal quantity of *spirits* of *Harts-horn, per se, i. e.* not made with Lime, did in 4 hours absorb 3 or 4 cubick inches of air; and the following day it remitted or generated two cubick inches of air: The third day turning from very warm to cold, it again resorbed that air, and continued in an absorbing state for a day or two.

That there is great plenty of air incorporated into the substance of *Vegetables*, which by the action of fermentation is rouzed into an elastick state, is evident by these following Experiments, *viz.*

Experiment LXXXV.

March the 2d, I poured into the bolthead *b* (Fig. 34.) forty two cubick inches of *Ale* from the Tun, which had been there set to ferment 34 hours before: From that time to the 9th of *June* it generated 639 cubick inches of air, with a very unequal progression, more or less as the weather was warm, cool, or cold, and sometimes upon a change from warm to cool, it resorbed air, in all 32 cubick inches.

Experiment LXXXVI.

March the 2d, 12 cubick inches of *Malaga Raisins*, with 18 cubick inches of *water* generated by the 16th of *April* 411 cubick inches of air; and then in 2 or 3 cold days it resorbed 35 cubick inches. From the 21st of *April* to the 16th of *May* it generated 78 cubick inches; after which to the 9th of *June* it continued in a resorbing state, so as to resorb 13 cubick inches; there were at this season many hot days, with much Thunder and Lightning, which destroys the air's elasticity; so there was generated in all 489 cubick inches, of which 48 were resorbed. The liquor was at last very vapid.

From the great quantity of air generated from *Apples*, in the following Experiment, 'tis probable, that much more air would have risen from the laxer texture of ripe undryed Grapes, than did from these Raisins.

We see from these Experiments on Raisins and Ale, that in warm weather Wine and Ale do not turn vapid by imbibing air, but by fermenting and generating too much, whereby they are deprived of their enlivening principle, the air; for which reason these liquors are best preserved in cool cellars, whereby this active invigorating principle is kept within due bounds, which when they exceed, Wines are upon the fret and in danger of being spoiled.

Experiment LXXXVII.

Twenty-six cubick inches of *Apples* being mashed *August* 10, they did in 13 days generate 968 cubick inches of air, a quantity equal to 48 times their bulk; after which they did in 3 or 4 days resorb a quantity equal to their bulk, notwithstanding it was very hot weather; after which they were stationary, neither resorbing nor generating air in many days.

A very coarse *brown-sugar,* with an equal quantity of water, generated nine times its bulk of air; *Rice-flower* six times its bulk; *Scurvy-grass* leaves generated and absorbed air; Pease, Wheat and Barley did in Fermentation also generate great quantities of Air.

That this Air, which arises in such great quantities from fermenting and dissolving vegetables is true permanent Air, is certain, by its continuing in the same expanded elastick state for many weeks and months; which expanding watry vapours will not do, but soon condense when cool. And that this new generated air is elastical is plain, not only by its dilating and contracting with heat and cold, as common air does, but also by its being compressible, in proportion to the incumbent weight, as appears by the two following Experiments, which shew what the great force of these aerial particles is, at the instant they escape from the fermenting vegetables.

EXPERIMENT LXXXVIII.

I filled the strong *Hungary-water Bottle b c* Fig. 36. near half full of Pease, and then full of water, pouring in first half an inch depth of Mercury; then I screwed at *b* into the bottle the long slender tube *a z*, which reached down to the bottom of the bottle; the water was in two or three days all imbibed by the Pease, and they thereby much dilated; the Mercury was also forced up the slender glass tube near 80 inches high; in which state the new generated air in the bottle was compressed with a force equal to more than two Atmospheres and an half; if the bottle and tube were swung too and fro, the Mercury would make long vibrations in the tube between *z* and *b*, which proves the great elasticity of the compressed air in the bottle.

EXPERIMENT LXXXIX.

I found the like elastick force by the following Experiment, *viz.* I provided a strong iron pot *a b c d* Fig. 37. which was 2 and $\frac{1}{4}$ inches diameter within side, and five inches deep. I poured into it half an inch depth of Mercury; then I put a little coloured honey at *x*, into the bottom of the glass tube *z x*, which was sealed at the top. I set this tube in the iron cylinder *n n*, to save it from breaking by the swelling of the Pease. The pot being filled with Pease and water, I put a leathern collar between the mouth and lid of the pot, which were both ground even, and then pressed the lid hard down in a Cyder-press: The third day I opened the pot and found all the water imbibed by the Pease; the Honey was forced up the glass-tube by the Mercury to *z*, (for so far the glass was dawbed) by which means I found the pressure had been equal to two atmospheres and $\frac{1}{4}$; and the diameter of the pot being $2+\frac{3}{4}$ inches, its area was six square inches, whence the dilating force of the air against the lid of the pot was equal to 189 pounds.

And that the expansive force of new generated air is

vastly superior to the power with which it acted on the Mercury in these two Experiments is plain from the force with which fermenting Must will burst the strongest vessels; and from the vast explosive force with which the air generated from nitre in the firing of gun-powder, will burst asunder the strongest bombs or cannon, and whirl fortifications in the air.

This sort of mercurial gage, made use of in Experiment 89, with some unctuous matter, as Honey, Treacle, or the like, on the Mercury in the tube, to note how high it rises there, might probably be of service, in finding out unfathomable depths of the Sea, *viz.* by fixing this sea-gage to some buoyant body which should be sunk by a weight fixt to it, which weight might be an easy contrivance be detached from the buoyant body, as soon as it touched the bottom of the sea; so that the buoyant body and gage would immediately ascend to the surface of the water; the buoyant body ought to be pretty large, and much lighter than the water, that by its greater eminence above the water it might the better be seen; for 'tis probable that from great depths it may rise at a considerable distance from the ship, tho' in a calm.

For greater accuracy it will be needful, first to try this sea gage, at several different depths, down to the greatest depth that a line will reach, thereby to discover, whether or how much the spring of the air is disturbed or condensed, not only by the great pressure of the incumbent water, but also by its coldness at great depths; and in what proportion, at different known depths, and in different lengths of time, that an allowance may accordingly be made for it at unfathomable depths.

This gage will also readily shew the degrees of compression in the condensing engine.

But to return to the subject of the two last Experiments, which prove the elasticity of this new generated air; which elasticity is supposed to consist in the active aerial particles, repelling each other with a force, which is reciprocally proportional to their distances. That illustrious Philosopher, Sir *Isaac Newton*, in accounting how air and vapour is produced, Opticks *Quer*. 31. says, "The particles when they are shaken off from bodies by heat or fermentation so soon as they are beyond the reach of the attraction of the body receding from it, as also from one another, with great strength and keeping at a distance, so as sometimes to take up above a million of times more space than they did before in the form of a dense body, which vast contraction and expansion seems unintelligible, by feigning the particles of Air to be springy and ramous, or rolled up like hoops, or by any other means than by a repulsive power." The truth of which is further confirmed by these Experiments, which shew the great quantity of air emitted from fermenting bodies; which not only proves the great force with which the parts of those bodies must be distended; but shews also how very much the particles of air must be coiled up in that state, if they are, as has been supposed, springy and ramous.

To instance in the case of the pounded Apples which generated above 48 times their bulk of Air; this air, when in the Apples, must be compressed into less than a forty eighth part of the space it takes up, when freed from them, and it will consequently be 48 times more dense; and since the force of compressed air is proportional to its density, that force which compresses and confines this Air in the Apples, must be equal to the weight of 48 of our atmospheres, when the Mercury in the Barometer stands at fair, that is 30 inches high.

Now a cubick inch of Mercury weighing 3580 grains, thirty cubick inches (which is equal to the weight of our atmosphere on an area of a cubick inch) will weigh 15 pounds, 5 ounces, 215 grains; and 48 of them will weigh above 736 pounds; which is therefore equal to the force with which an inch square of the surface of the Apple would compress the air, supposing there were no other substance but air in the Apple: And if we take the surface of an Apple at 16 square inches, then the whole force with which that surface would compress the included air, would be 11776 pounds. And since action and re-action are equal, this would be the force, with which the air in the Apple would endeavour to expand itself, if it were there in an elastick and strongly compressed state: But so great an expansive force in an Apple would certainly rend the substance of it with a strong explosion, especially when that force was increased, by the vigorous influence of the Sun's warmth.

We may make a like estimate also, from the great quantities of air which arose either by fermentation, or the force of fire from several other bodies. Thus in Exp. 55. there arose from a piece of heart of *Oak*, 216 times its bulk of air. Now 216 cubick inches of air, compressed into the space of one cubick inch, would, if it continued there in an elastick state, press against one side of the cubick inch with an expansive force equal to 3310 pounds weight, supposing there were no other substance but air contained in it; and it would press against the six sides of the cube, with a force equal to 19860 pounds, a force sufficient to rend the *Oak* with a vast explosion: 'tis very reasonable therefore to conclude, that most of these now active particles of the new generated air, were in a fix'd state in the Apple and Oak before they were roused, and put into an active repelling state by fermentation and fire.

The weight of a cubick inch of Apple being 191 grains, the weight of a cubick inch of air $\frac{2}{7}$ of a grain, 48 times that weight of air is nearly equal to the fourteenth part of the weight of the Apple.

And if to the air thus generated from a vessel of any vegetable liquor by fermentation, we add the air that might afterwards be obtained from it, by heat or distillation; and to that also the vast quantity of air which by Experiment 73 is found to be contained in its *Tartar*, which adheres to the sides of the vessel; it would by this means be found that air makes a very considerable part of the substance of Vegetables, as well as of Animals.

But tho' from what has been said, it is reasonable to think, that many of these particles of air were in a fixt state, strongly adhering to and wrought into the substance of Apples; yet on the other hand it is most evident from Exper. 34 and 38, where innumerable bubbles of air incessantly arose through the sap of Vines, that there is a considerable quantity of air in Vegetables, upon the wing, and in a very active state, especially in warm weather, which enlarges the sphere of their activity.

The effects of the Fermentation of mineral Substances on the Air.

I have above shewn that air may be produced from mineral Substances, by the action of fire in distillation. And we have in the following Experiments many instances of the great plenty of air which is generated by some fermenting mixtures, absorbed by others, and by others alternately generated and absorbed.

EXPERIMENT XC.

I poured upon a middle sized *Gold Ring*, beat into a thin plate, two cubick inches of *Aqua Regia*; the *Gold* was all dissolved the next day, when I found 4 cubick inches of air generated; for air bubbles were continually arising during the solution: But since *Gold* loses nothing of its weight in being thus dissolved, the 4 cubick inches of air, which weighed more than a grain, must arise either out of the pores of the *Gold*, or from the *Aqua Regia*, which makes it probable, that there are air particles in acid spirits; for by Experiment 75, they absorb air, which air particles regained their elasticity, when the acid spirits which adhered to them were more strongly attracted by the gold, than by the air particles.

Experiment XCI.

A quarter of a cubick inch of *Antimony*, and two cubick inches of *Aqua-regia*, generated 38 cubick inches of air, the first 3 or 4 hours, and then absorbed 14 cubick inches in an hour or two; after which it was stationary, till I let into the glass vessel *a y* (Fig. 34.) about a quart of fresh air: Upon which it absorbed so fast, as to make the water rise very visibly in *a y*, whereby it absorbed 30 cubick inches more. It is very observable, that air was generated while the ferment was small, on the first mixing of the ingredients: But when the ferment was greatly increased, so that the fumes rose very visibly, then there was a change made from a generating to an absorbing state; that is, there was more air absorbed than generated.

That I might find whether the air was absorbed by the fumes only of the *Aqua-regia*, or by the acid sulphureous vapours, which ascended from the *Antimony*, I put a like quantity of *Aqua-regia* into a bolthead *b*, (Fig. 34.) and heated it by pouring a large quantity of hot water into the cistern *x x*, which stood in a larger vessel, that retained the hot water about it, but no air was absorbed; for when all was cold, the water stood at the point *z*, where I first placed it: Yet in the distillation of compound Aqua-fortis, Exper. 75. a little was absorbed. Hence therefore it is probable, that the greatest part, if not all the air, was absorbed by the fumes, which arose from the *Antimony*.

Experiment XCII.

Some time in *February*, the weather very cold, I poured upon a quarter of cubick inch of powdered *Antimony*, a cubick inch of *compound or double Aqua-fortis*, in the bolthead *b*, (Fig. 34.) in the first 20 hours it generated about 8 cubick inches of air; after that, the weather being somewhat warmer, it fermented faster, so as in 2 or 3 hours to generate 82 cubick inches of air more; but the following night being very cold, little was generated: So the next morning I poured hot water into the vessel *x x*, which renewed the ferment, so that it generated 4 cubick inches more, in all 130 cubick inches, a quantity equal to 520 times the bulk of the *Antimony*.

The fermented mass looked like Brimstone, and when heated over the fire, there sublimed into the neck of the bolthead a red sulphur, and below it a yellow, which sulphur, as Mr. *Boyle* observes, Vol. III, *p*. 272. cannot be obtained by the bare action of fire, without being first well digested in oil of Vitriol, or spirit of Nitre. And by comparing the quantity of air obtained by fermentation in this Experiment, with the quantity obtained by the force of fire in Exper. 69. we find that five times more air was generated by fermentation than by fire, which shews fermentation to be a more subtile dissolvent than fire; yet in some cases there is more air generated by fire than by fermentation.

Half a cubick inch of *oil* of *Antimony*, with an equal quantity of compound *Aqua-fortis*, generated 36 cubick inches of elastick air, which was all resorbed the following day.

EXPERIMENT XCIII.

Some time in *February*, a quarter of a cubick inch of *filings* of *Iron*, and a cubick inch of *compound Aqua-fortis*, without any water, did in 4 days absorb 27 cubick inches of air. It having ceased to absorb, I poured hot water into the vessel *x x*, to try if I could renew the ferment. The effect of this was, that it generated 3 or 4 cubick inches of air, which continued in that state for some days, and was then again resorbed.

I repeated the same Experiment in warm weather in *April*, when it more briskly absorbed 12 cubick inches in an hour.

Experiment XCIV.

March 12th, $\frac{1}{4}$ of a cubick inch of *filings* of *Iron*, with a cubick inch of *compound Aqua-fortis*, and an equal quantity of water, for the first half hour absorbed 5 or 6 cubick inches of air; but in an hour more it had emitted that quantity of air; and in two hours more it again resorbed what had been just before emitted. The day following it continued absorbing, in all 12 cubick inches: And then remained stationary for 15 or 20 hours. The third day it had again remitted or generated 3 or 4 cubick inches of air, and thence continued stationary for five or six days.

A like quantity of *filings* of *Iron*, and *oil* of *Vitriol*, made no sensible ferment, and generated a very little air; but upon pouring in an equal quantity of water, it generated in 21 days 43 cubick inches of air; and in 3 or 4 days more it resorbed 3 cubick inches of air; when the weather turned warmer it was generated again, which was again resorbed when it grew cool.

$\frac{1}{4}$th Of a cubick inch of *filings* of *Iron*, and a cubick inch of *oil* of *Vitriol*, with three times its quantity of *Water*, generated 108 cubick inches of air.

Filings of *Iron*, with *spirit* of *Nitre*, either with an equal quantity of *water*, or without *water*, absorbed air, but most without water.

$\frac{1}{4}$th Of a cubick inch of *filings* of *Iron*, and a cubick inch of *Limon juice*, absorbed two cubick inches of air.

It is remarkable, that the same mixtures should change from generating to absorbing, and from absorbing to generating states; sometimes with, and sometimes without any sensible alteration of the temperature of the air.

Experiment XCV.

Half a cubick inch of *spirits* of *Harts-horn*, with *filings* of *Iron* absorbed $1+\frac{1}{2}$ cubick inches of air, with *filings* of

Copper double that quantity of air, and made a very deep blue tincture, which it retained long, when exposed to the open air. It was the same with *spirit* of *Sal Ammoniac*, and *filings* of *Copper*.

A quarter of a cubick inch of *filings* of *Iron*, with a cubick inch of powdered *Brimstone*, made into a paste with a little water, absorbed 19 cubick inches of air in two days. *N. B.* I poured hot water into the cistern *x x*, (Fig. 34.) to promote the ferment.

A like quantity of *filings* of *Iron*, and powdered *Newcastle Coal*, did in 3 or 4 days generate 7 cubick inches of air. I could not perceive any sensible warmth in this mixture, as was in the mixture of *Iron* and *Brimstone*.

Powdered Brimstone and *Newcastle Coal* neither generated nor absorbed.

Filings of Iron and *Water* absorbed 3 or 4 cubick inches of air, but they do not absorb so much when immersed deep in water; what they absorb is usually the first 3 or 4 days.

Filings of *Iron*, and the abovementioned *Walton Pyrites* in Exper. 70. absorbed in 4 days a quantity of air nearly equal to double their bulk.

Copper Oar, and *compound Aqua-fortis*, neither generated nor absorbed air, but mixed with *water* it absorbed air.

A quarter of a cubick inch of *Tin*, and double that quantity of *compound Aqua-fortis*, generated two cubick inches of air; part of the *Tin* was dissolved into a very white substance.

Experiment XCVI.

April 16th, A cubick inch of the aforementioned *Walton Pyrites* powder'd, with a cubick inch of *compound Aqua-fortis*, expanded with great violence, heat and fume into a space equal to 200 cubick inches, and in a little time it condensed into its former space, and then absorbed 85 cubick inches of air.

But the like quantity of the same *Mineral*, with equal quantities of *compound Aqua-fortis* and *Water*, fermented more violently and generated above 80 cubick inches of air. I repeated these Experiments several times, both with and without *water*, and found constantly the same effect.

Yet *oil* of *Vitriol* and *Water*, with some of the same *Mineral*, absorbed air. It was very warm, but did not make a great ebullition.

Experiment XCVII.

I chose two equal sized boltheads, and put into each of them a cubick inch of powdered *Walton Pyrites*, with only a cubick inch of *compound Aqua-fortis* into one, and a cubick inch of *Water* and *compound Aqua-fortis* into the other: Upon weighing all the ingredients and vessels exactly, both before and after the fermentation, I found the bolthead with *compound Aqua-fortis* alone had lost in fumes 1 dram 5 grains: But the other bolthead with *Water and compound Aqua-fortis*, which fumed much more, had lost 7 drams, 1 scruple, 7 grains, which is six times as much as the other lost.

Experiment XCVIII.

A cubick inch of *Newcastle Coal* powdered, and an equal quantity of *compound Aqua-fortis* poured on it, did in 3 days absorb 18 cubick inches of air; and in 3 days more it remitted and generated 12 cubick inches of air; and on pouring warm *water* into the vessel *x x* (Fig. 34.) it remitted all that had been absorbed.

Equal quantities of *Brimstone* and *compound Aqua-fortis* neither generated nor absorbed any air, notwithstanding hot *water* was poured into the vessel *x x*.

A cubick inch of finely powdered *Flint*, and an equal quantity of *compound Aqua-fortis*, absorbed in 5 or 6 days 12 cubick inches of air.

Equal quantities of powdered *Bristol Diamond*, and *compound Aqua-fortis*, and *Water* absorbed 16 times their bulk of air.

The like quantities without Water absorbed more slowly 7 times their bulk of air.

Powdered *Bristol Marble* (*viz.* the shell in which those *Diamonds* lay) covered pretty deep with *water*, neither generated nor absorbed air; and it is well known that *Bristol water* does not sparkle like some other *Mineral waters*.

Experiment XCIX.

When *Aqua-regia* was poured on *Oleum Tartari per Deliquium* much air was generated, and that probably chiefly from the *Oleum Tartari*; for by Exper. 74. *Sal Tartar* has plenty of air in it.

It was the same when *oil* of *Vitriol* was poured on *Ol. Tartari;* and *Ol. Tartari* dropped on boyling Tartar generated much air.

When equal quantities of *Water* and *oil* of *Vitriol* were poured on the sea salt it absorbed 15 cubick inches of air; but when in the like mixture the quantity of *Water* was double to that of the *oil* of *Vitriol*, then but half so much air was absorbed.

Experiment C.

I will next shew, what effects several *Alkaline Mineral* bodies had on the air in fermenting mixtures.

A solid cubick inch of unpowdered *Chalk*, with an equal quantity of *oil* of *Vitriol*, fermented much at first, and in some degree for 3 days; they generated 31 cubick inches of air. The *Chalk* was only a little dissolved on its surface.

Yet *Lime* made of the same *Chalk* absorbed much air, when *oil of Vitriol* was poured on it, and the ferment was so violent that it breaking the glass vessels, I was obliged to put the ingredients in an *Iron* vessel.

Two cubick inches of fresh *Lime*, and four of common *white wine Vinegar* absorbed in 15 days 22 cubick inches of air.

The like quantity of fresh *Lime* and *Water* absorbed in 3 days 10 cubick inches of air.

Two cubick inches of *Lime*, and an equal quantity of *Sal Ammoniac* absorbed 115 cubick inches.

A quart of unslaked *Lime*, left for 44 days, to slaken gradually by it self without any mixture, absorbed no air.

March 3d, a cubick inch of powdered *Belemnitis*, taken from a *Chalk* pit, and an equal quantity of *oil* of *Vitriol*, generated in 5 minutes 35 cubick inches of air. *March* 5th, it had generated 70 more. *March* 6th, it being a hard frost, it resorbed 12 cubick inches; so it generated in all 105 inches, and resorbed 12.

Powdered *Belemnitis* and *Limon juice* generated plenty of air too; as did also the *Star-Stone, Lapis Judaicus*, and *Selenitis* with *oil* of *Vitriol.*

Experiment CI.

Gravelled, that is well burnt, *Wood-ashes, decrepitated Salt*, and *Colcothar* of *Vitriol*, placed severally under the inverted glass *z z a a* (Fig. 35.) increased in weight by imbibing the floating moisture of the air: But they absorbed no elastick air. It was the same with the remaining *lixivious Salt* of a distillation of *Nitre.*

But 4 or 5 cubick inches of powdered fresh *Cynder* of *Newcastle Coal* did in seven days absorb 5 cubick inches of elastick air. And 13 cubick inches of air were in 5 days absorbed by *Pulvis Urens*, a powder which immediately kindles into a live Cole, upon being exposed to the open air.

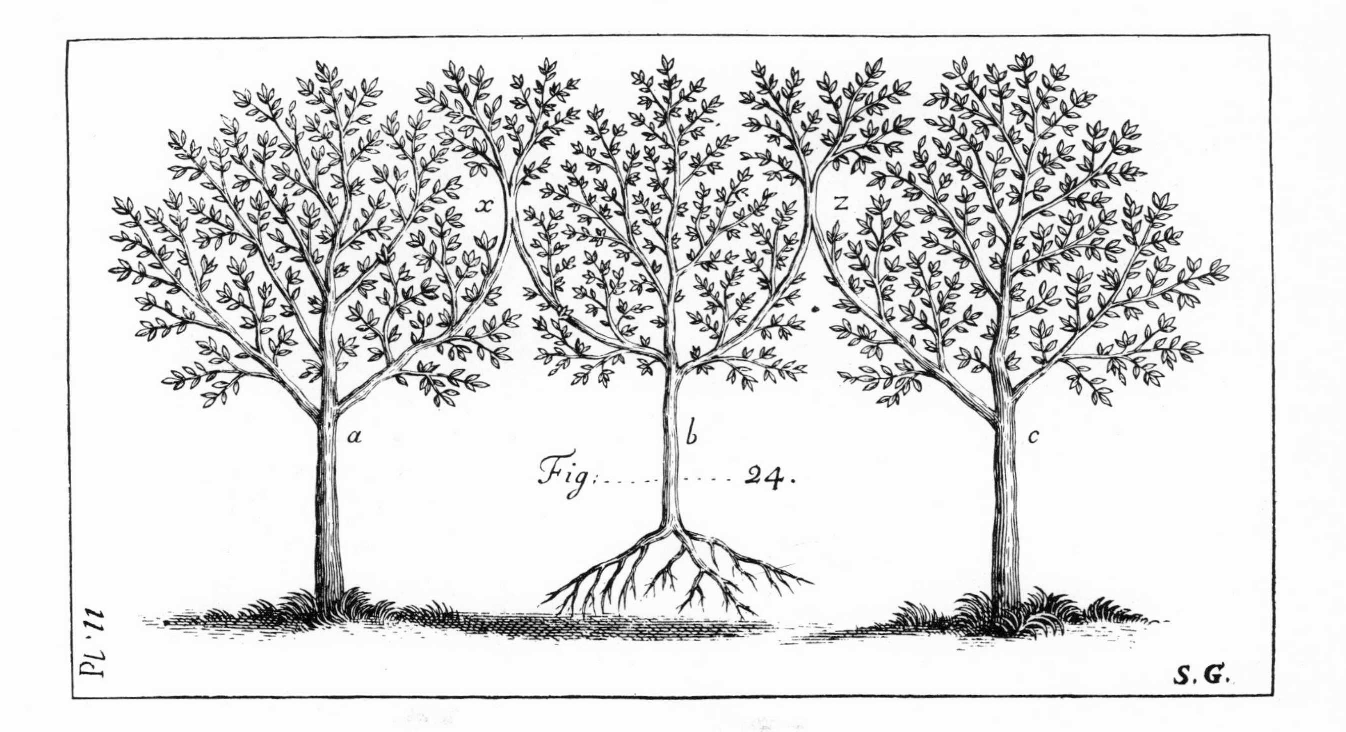
x
z
a
b
c
Fig. 24.
Pl. 11
S.G.

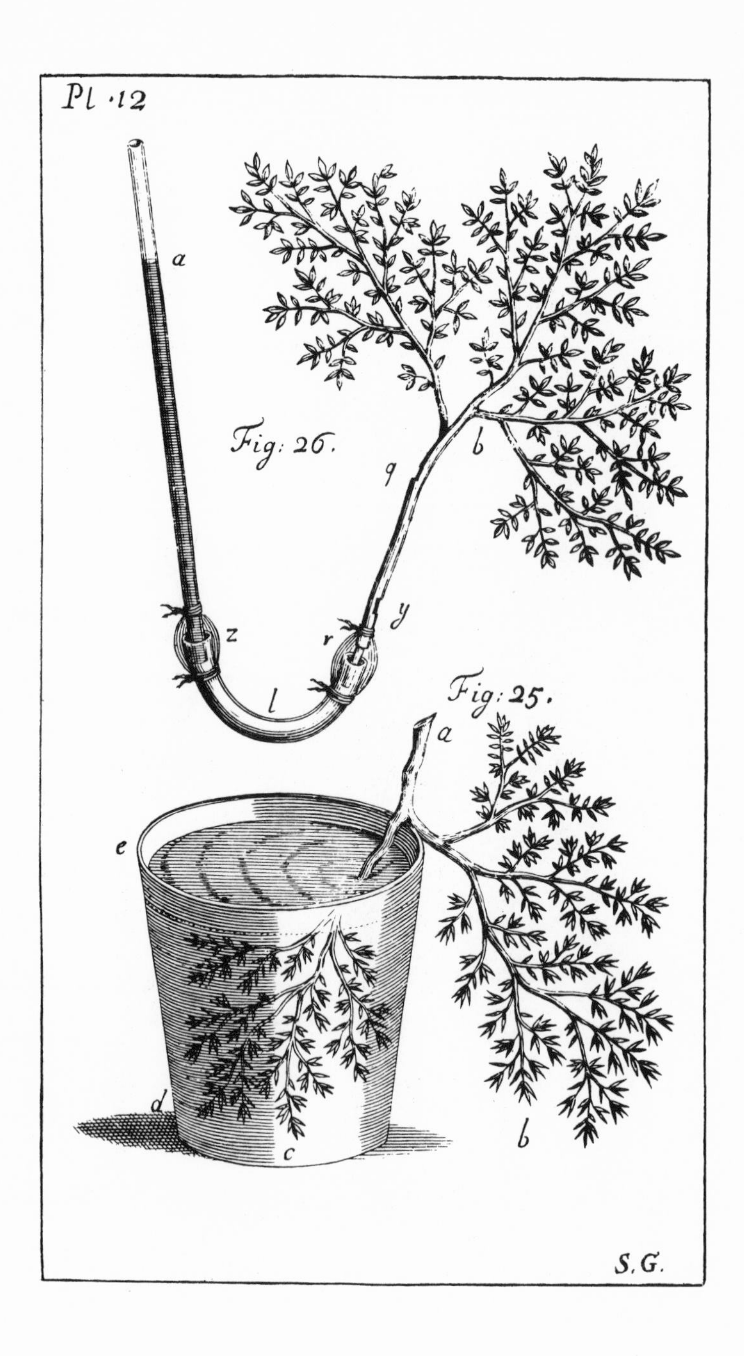
Pl ·12
a
Fig: 26.
q
b
z
r
y
l
Fig: 25.
a
e
d
c
b
S.G.

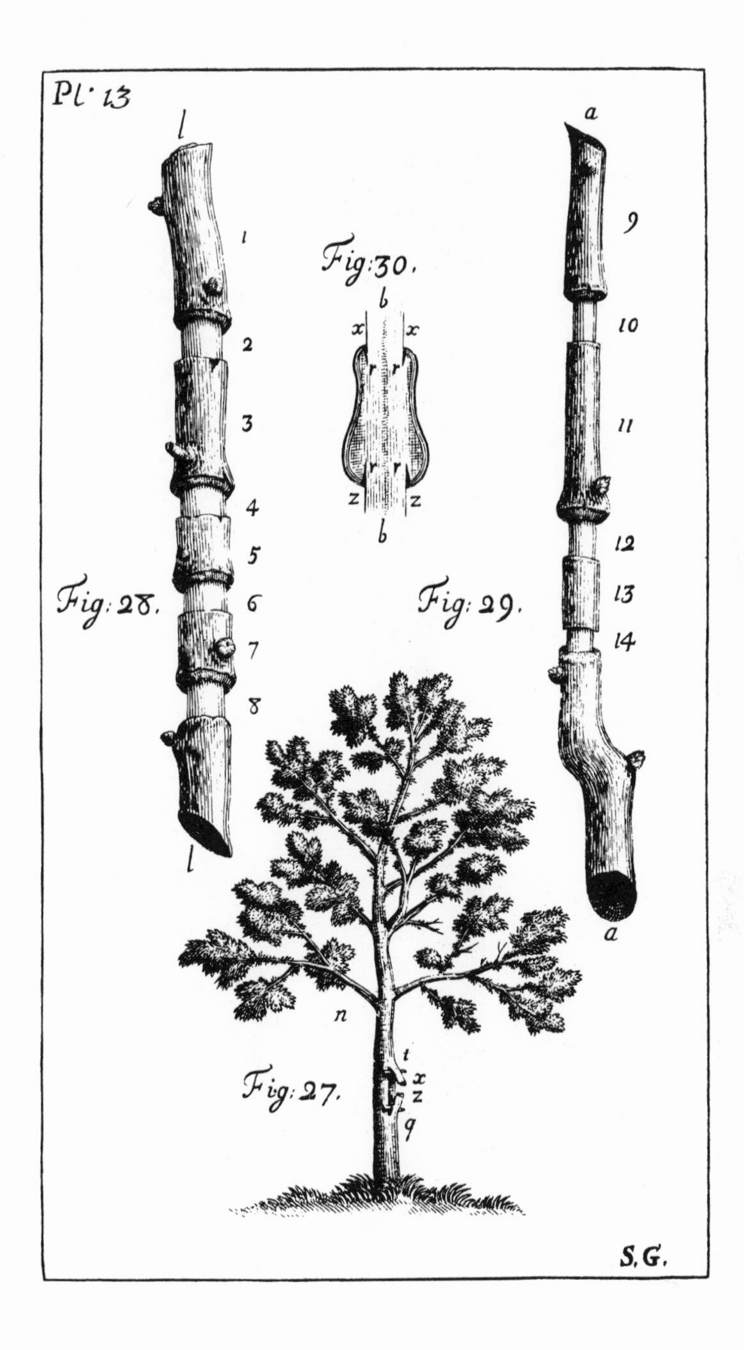

Pl. 13
l
1
2
3
4
5
6
7
8
l
Fig: 28.
Fig: 30.
b
x x
r r
r r
z z
b
a
9
10
11
12
13
14
a
Fig: 29.
n
t
x
z
q
Fig: 27.
S.G.

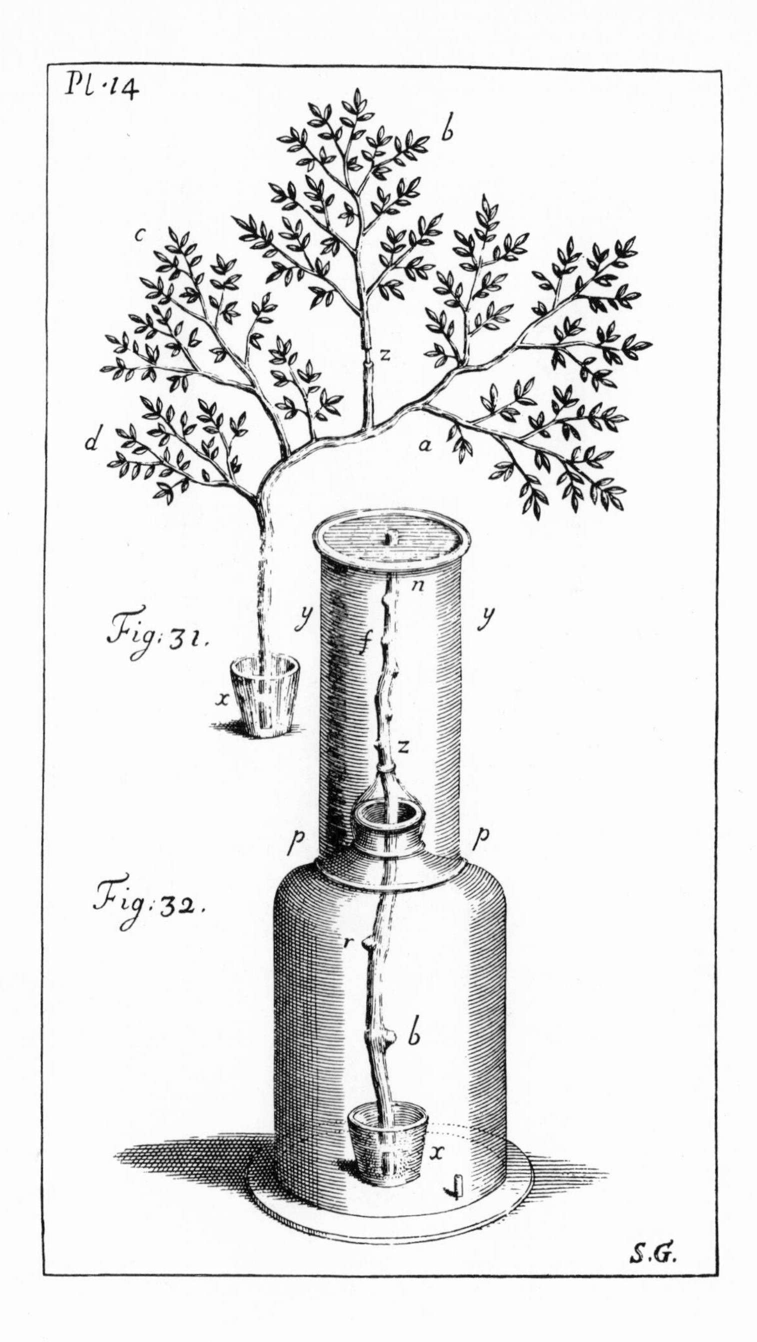
Pl·14
b
c
z
d
a
n
y
y
Fig: 31.
f
x
z
p
p
Fig: 32.
r
b
x
S.G.

EXPERIMENT CII.

What effect burning and flaming bodies, and the respiration of Animals have on the air, we shall see in the following Experiments, *viz.*

I fix'd upon the pedestal under the inverted glass *z z a a*, (Fig. 35.) a piece of *Brown Paper*, which had been dipped in a solution of *Nitre*, and then well dryed; I set fire to the Paper by means of a burning glass: The *Nitre* detonized and burnt briskly for some time, till the glass *z z a a* was very full of thick fumes, which extinguished it. The expansion caused by the burning *Nitre*, was equal to more than two quarts: When all was cool, there was near 80 cubick inches of new generated air, which arose from a small quantity of detonized *Nitre*; but the elasticity of this new air daily decreased, in the same manner as Mr. *Hauksbee* observed the air of fired Gun-powder to do, *Physico-mechanical Exper. p.* 83. so that he found 19 of 20 parts occupied by this air to be deserted in 18 days, and its space filled by the ascending water; at which station it rested, continuing there for 8 days without alteration: And in like manner, I found that a considerable part of the air which was produced by fire in the distillation of several substances, did gradually lose its elasticity in a few days after the distillation was over; but it was not so when I distilled air thro' water, as in Experiment 77. (Fig. 38.)

EXPERIMENT CIII.

I placed on the same pedestal large *Matches* made of linen rags dipped in melted *Brimstone*: The capacity of the vessel, (Fig. 35.) above *z z* the surface of the water, was equal to 2024 cubick inches. The quantity of air which was absorbed by the burning *Match* was 198 cubick inches, equal to $\frac{1}{10}$ part of the whole air in the vessel.

I made the same Experiment in a lesser vessel *z z a a*,

(Fig. 35.) which contained but 594 cubick inches of air, in which 150 cubick inches were absorbed; *i. e.* full $\frac{1}{4}$ part of the whole air in the receiver: So that tho' more air is absorbed by burning *Matches* in large vessels, where they burn longest, than in small ones, yet more air, in proportion to the bulk of the vessel, is absorbed in small than in large vessels: If a fresh *Match* were lighted, and put into this infected air, tho' it would not burn $\frac{1}{5}$ part of the time that the former *Match* burnt in fresh untainted air, yet it would absorb near as much air in that short time; and it was the same with Candles.

Experiment CIV.

Equal quantities of *filings* of *Iron* and *Brimstone*, when let fall on a hot Iron on the pedestal under the inverted glass *z z a a*, (Fig. 35.) did in burning absorb much air; and it was the same with *Antimony* and *Brimstone*: Whence it is probable, that *Vulcano's*, whose fewel consist chiefly of *Brimstone*, mix'd with several mineral and metaline substances, do not generate, but rather absorb air.

We find in the foregoing Experiment 102 on *Nitre*, that a great part of the new generated air is in a few days resorbed, or loses its elasticity: But the air which is absorbed by burning *Brimstone*, or the flame of a Candle, does not recover its elasticity again, at least, not while confined in my glasses.

Experiment CV.

I made several attempts to try whether air full of the fumes of burning *Brimstone* was as compressible as common fresh air, by compressing at the same time tubes full of each of these airs in the condensing engine; and I found that clear air is very little more compressible than air with fumes of *Brimstone* in it: But I could not come to an exact certainty in the matter, because the fumes were at the same time destroying the elasticity of the air. I took care to make the air in both tubes of the same temperature, by first immersing them in cold water, before I compressed them.

Experiment CVI.

I set a lighted *tallow Candle*, which was about $\frac{6}{10}$ of an inch diameter, under the inverted receiver *z z a a*, (Fig. 35.) and with a syphon I immediately drew the water up to *z z*: Then drawing out the syphon, the water would descend for a quarter of a minute, and after that ascend, notwithstanding the Candle continued burning, and heating the air for near 3 minutes. It was observable in this Experiment, that the surface of the water *z z* did not ascend with an equal progression, but would be sometimes stationary; and it would sometimes move with a slow, and sometimes with an accelerated motion; but the denser the fumes the faster it ascended. As soon as the Candle was out, I marked the height of the water above *z z*, which difference was equal to the quantity of air, whose elasticity was destroyed by the burning Candle. As the air cooled and condensed in the receiver, the water would continue rising above that mark, not only till all was cool, but for 20 or 30 hours after that, which height it kept, tho' it stood many days; which shews that the air did not recover the elasticity which it had lost.

The event was the same, when for greater accuracy I repeated this Experiment by lighting the Candle after it was placed under the receiver, by means of a burning-glass, which set fire to a small piece of brown paper fixed to the wick of a Candle, which paper had been first dipped in a strong solution of *Nitre* in Water, and when well dried, part of it was dipped in melted *Brimstone*; it will also light the Candle without being dipped in *Brimstone*. Dr. *Mayow*, *found* the bulk of the air lessened by $\frac{1}{30}$ part, but does not mention the size of the glass vessel under which he put the lighted Candle, *De Sp. Nitro-aereo*, *p.* 101. The capacity of the vessel above *z z*, in which the Candle burnt in my Experiment, was equal to 2024 cubick inches; and the elasticity of the $\frac{1}{26}$ part of this air was destroyed.

The Candle cannot be lighted again in this infected air by a burning-glass: But if I first lighted it, and then put it into the same infected air, tho' it was extinguished in $\frac{1}{5}$ part of the time, that it would burn in the same vessel, full of fresh air; yet it would destroy the elasticity of near as much air in that short time, as it did in five times that space of time in fresh air; this I repeated several times, and found the same event: Hence a gross air which is loaded with vapours, is more apt in equal times to lose its elasticity in greater quantities, than a clear air.

I observe that where the vessels are equal, and the size of the Candles unequal, the elasticity of more air will be destroyed by the large than by the small Candle: And where Candles are equal, there most air in proportion to the bulk of the vessel will be absorbed in the smallest vessel: Tho' with equal Candles there is always most elastick air destroyed in the largest vessel, where the Candle burns longest.

I found also in fermenting liquors, that *cæteris paribus*, more air was either generated or absorbed in large, than in small vessels, by generating or absorbing mixtures. As in the mixture of *Aqua Regia* and *Antimony* in Experiment 91, by enlarging the bulk of the air in the vessel, a greater quantity of air was absorbed. Thus also *filings* of *Iron* and *Brimstone*, which in a more capacious vessel absorbed 19 cubick inches of air, absorbed very little when the bulk of air above the ingredients was but 3 or 4 cubick inches: For I have often observed, that when any quantity of air is saturated with absorbing vapours to a certain degree, then no more elastick air is absorbed: Notwithstanding the same quantity of absorbing substances would, in a larger quantity of air, have absorbed much more air; and this is the reason why I was never able to destroy the whole elasticity of any included bulk of air, whether it was common air, or new generated air.

EXPERIMENT CVII.

May 18, which was a very hot day, I repeated Dr. *Mayow*'s Experiment, to find how much air is absorbed by the breath of Animals inclosed in glasses, which he found with a mouse to be $\frac{1}{14}$ part of the whole air in the glass vessel *De Sp. Nitro-aereo, p.* 104.

I placed on the pedestal, under the inverted glass *z z a a*, (Fig. 35.) a full grown *Rat*. At first the water subsided a little, which was occasioned by the rarifaction of the air, caused by the heat of the Animal's body. But after a few minutes the water began to rise, and continued rising as long as the Rat lived, which was about 14 hours. The bulk of the air in which the Rat lived so many hours was 2024 cubick inches; the quantity of elastick air which was absorbed was 73 cubick inches, above $\frac{1}{27}$ part of the whole, nearly what was absorbed by a Candle in the same vessel, in Experiment 106.

I placed at the same time in the same manner another almost half grown *Rat* under a vessel, whose capacity above the surface of the water *z z* (Fig. 35.) was but 594 cubick inches, in which it lived 10 hours; the quantity of elastick air which was absorbed, was equal to 45 cubick inches, *viz.* $\frac{1}{13}$ part of the whole air, which the *Rat* breathed in: A *Cat* of 3 months old lived an hour in the same receiver, and absorbed 16 cubick inches of air, *viz.* $\frac{1}{30}$ part of the whole; an allowance being made in this estimate for the bulk of the Cat's body. A candle in the same vessel continued burning but one minute, and absorbed 54 cubick inches, $\frac{1}{11}$ part of whole air.

And as in the case of burning *Brimstone* and *Candles*, more air was found to be absorbed in large vessels, than in small ones; and *vice versâ*, more Air in proportion to the capacity of the vessel was absorbed in small, than in large vessels, so the same holds true here too in the case of animals.

Experiment CVIII.

The following Experiment will shew, that the elasticity of the air is greatly destroyed by the *respiration of human lungs*, viz.

I made a bladder very supple by wetting of it, and then cut off so much of the neck as would make a hole wide enough for the biggest end of a large fosset to enter, to which the bladder was bound fast. The bladder and fosset contained 74 cubick inches. Having blown up the bladder, I put the small end of the fosset into my mouth; and at the same time pinched my nostrils close that no air might pass that way, so that I could only breath to and fro the air contained in the bladder. In less than half a minute I found a considerable difficulty in breathing, and was forced after that to fetch my breath very fast; and at the end of the minute, the suffocating uneasiness was so great that I was forced to take away the bladder from my mouth. Towards the end of the minute the bladder was become so flaccid, that I could not blow it above half full with the greatest expiration that I could make: And at the same time I could plainly perceive that my lungs were much fallen, just in the same manner as when we breathe out of them all the air we can at once. Whence it is plain that a considerable quantity of the elasticity of the air contained in my lungs, and in the bladder was destroyed: Which supposing it to be 20 cubick inches, it will be $\frac{1}{13}$ part of the whole Air, which I breathed to and fro; for the bladder contained 74 cubick inches, and the lungs, by the following Experiment, about 166 cubick inches, in all 240.

These effects of respiration on the elasticity of the air, put me upon making an attempt to measure the inward surface of the lungs, which by a wonderful artifice are admirably contrived by the divine artificer, so as to make their inward surface to be commensurate to an expanse of air many times greater than the animal's body; as will appear from the following estimate, *viz.*

EXPERIMENT CIX.

I took the lungs of a Calf and cut off the heart and wind-pipe an inch above its branching into the lungs; I got nearly the specifick gravity of the substance of the lungs, (which is a continuation of the branchings of the wind-pipe, and blood-vessels) by finding the specifick gravity of the wind-pipe, which I had cut off; it was to Well-water as 1.05 to 1. And a cubick inch of water weighing 254 grains; I thence found by weighing the lungs the whole of their solid substance to be equal to $37+\frac{1}{2}$ cubick inches.

I then filled a large earthen vessel brim full of water, and put the lungs in, which I blew up keeping them under water with a pewter plate. Then taking the lungs out and letting the plate drop to the bottom of the water, I poured in a known quantity of water, till the vessel was brimful again; that water was 7 pounds 6 ounces and $\frac{1}{2}$ equal to 204 cubick inches; from which deducting the space occupied by the solid substance of the lungs, *viz.* $37+\frac{1}{2}$ cubick inches, there remains $166+\frac{1}{2}$ cubick inches for the cavity of the lungs. But as the Pulmonary Veins, Arteries and Lymphaticks will, when they are in a natural state replete with blood and lymph, occupy more space than they do in their present empty state; therefore some allowance must also be made, out of the above taken cavity of the lungs, for the bulk of those fluids; for which $25+\frac{1}{2}$ cubick inches seems to be a sufficient proportion, out of the $166+\frac{1}{2}$ cubick inches; so there remains 141 cubick inches for the cavity of the lungs.

I poured as much water into the *Bronchiæ* as they would take in, which was 1 pound 8 ounces, equal to 41 cubick inches; this deducted from the above found cavity of the lungs, there remains 100 cubick inches for the sum of the cavity of the vesicles.

Upon viewing some of these vesicles with a microscope, a middle sized one seems to be about $\frac{1}{100}$ part of an inch diameter; then the sum of the surfaces in a cubick inch of these small vesicles (supposing them to be so many little cubes, for they are not spherical) will be 600 square inches; for if the number of the divisions of the side of the cubick inch be 100, there will be 100 planes, containing each one square inch, in each dimension of the cube, which having three dimensions, the sum of those planes will be 300 square inches, and the sum of the surfaces of each side of those planes will be 600 square inches; which multiplied by the sum of all the vesicles in the lungs, *viz.* 100 cubick inches, will produce 60000 square inches; one third of which must be deducted, to make an allowance for the absence of two sides in each little vesicular cube, that there might be a free communication among them for the air to pass to and fro; so there remains 40000 square inches for the sum of the surface of all the vesicles.

And the *Bronchiæ* containing 41 cubick inches, supposing them at a medium to be cylinders of $\frac{1}{10}$ of an inch diameter, their surface will be 1635 square inches, which added to the surface of the vesicles makes the sum of the surface of the whole lungs to be 41635 square inches, or 289 square feet, which is equal to 10 times the surface of a man's body, which at a medium is computed to be equal to 15 square feet.

I have not had an opportunity to take in the same manner the capacity and dimensions of human lungs; the bulk of which Dr. *James Keill*, in his *Tentamina Medico-physica, p.* 80, found to be equal to 226 cubick inches. Whence he estimated the sum of the surface of the vesicles to be 21906 square inches. But the bulk of human lungs is much more capacious than 226 cubick inches; for Dr. *Jurin*, by an accurate Experiment, found that he breathed out, at one large expiration, 220 cubick inches of air; and I found it nearly the same, when I repeated the like Experiment in another manner: So that there must be a large allowance made for the bulk of the remaining air, which could not be expired from the lungs; and also for the substance of the lungs.

Supposing then, that according to Dr. *Jurin*'s estimate (in *Mott*'s *Abridgment of the Philosophical Transact.* Vol. I, p. 415.) we draw in at each common inspiration 40 cubick inches of air, that will be 48000 cubick inches in an hour, at the rate of 20 inspirations in a minute. A considerable part of the elasticity of which air is, we see by the foregoing Experiment, constantly destroyed, and that chiefly among the vesicles, where it is charged with much vapour.

But it is not easie to determine how much is destroyed. I attempted to find it out by the following Experiment, which I shall here give an account of, tho' it did not succeed so well as I could have wished, for want of much larger vessels; for if it was repeated with more capacious vessels, it would determine the matter pretty accurately; because by this artifice fresh air is drawn into the lungs at every inspiration, as well as in the free open air.

Experiment CX.

I made use of the syphon (Fig. 39.) taking away the bladders, and diaphragms *i i n n o*: I fixed by means of a bladder one end of a short leaden syphon to the lateral fosset *i i*: Then I fastened the large syphon in a vessel, and filled it with water, till it rose within two inches of *a*, and covered the other open end of the short syphon, which was depressed for that purpose. Over this orifice I placed a large inverted chymical receiver full of water; and over the other leg *o s* of the great syphon, I whelmed another large empty receiver, whose capacity was equal to 1224 cubick inches; the mouth of the receiver being immersed in the water, and gradually let down lower and lower by an assistant, as the water ascended in it. Then stopping my nostrils, I drew in breath at *a*, thro' the syphon from the empty receiver: And when that breath was expired, the valve *b i* stopping its return down thro' the syphon, it was forced thro' the valve *r*, and thence thro' the small leaden syphon into the inverted receiver full of water, which water descended as the breath ascended. In this manner I drew all the air, except 5 or 6 cubick inches, out of the empty receiver at *o*, the water at the same time ascending into it and filling it; by which means all the air in the empty receiver, as also all the air in the syphon *o s b*, was inspired into my lungs, and breathed out thro' the valve *r* into the receiver, which was at first full of water. I marked the boundary of air and water, and then immersed the whole receiver, which had the breath in it, under water, and there gradually poured the contained breath up into the other full receiver, which stood inverted over *o s*; whereby I could readily find whether the air had lost any of its elasticity: And for greater surety, I also measured the bulk of breath by filling the receiver with a known quantity of water up to the above mentioned mark; making also due allowance for a bulk of air, equal to the capacity of the large syphon *o s b*, which was at last sucked full of water.

The event was, that there was 18 cubick inches of air wanting; but as these receivers were much too small to make the Experiment with accuracy; that some allowance may be made for errors, I will set the loss of elastick air at 9 cubick inches, which is but $\frac{1}{136}$ part of the whole air respired, which will amount to 353 cubick inches in one hour, or 100 grains, at the rate of 48000 cubick inches inspired in an hour, or one ounce and a half in twenty four hours.

By pouring the like quantity of air to and fro under

water, I found that little or none of it was lost; so it was not absorbed by the water: To make this tryal accurately, the air must be detained some time under water, to bring it first to the same temperature with the water. Care also must be taken in making this Experiment, that the lungs be in the same degree of contraction, at the last breathing, as at the first, else a considerable error may arise from thence.

But tho' this be not an exact estimate, yet it is evident from the foregoing Experiments on respiration, that some of the elasticity of the air which is inspired is destroyed; and that chiefly among the vesicles, where it is most loaded with vapours; whence probably some of it, together with the acid spirits, with which the air abounds, are conveyed to the blood, which we see is by an admirable contrivance there spread into a vast expanse, commensurate to a very large surface of air, from which it is parted by very thin partitions; so very thin, as thereby probably to admit the blood and air particles (which are there continually changing from an elastick to a strongly attracting state) within the reach of each other's attraction, whereby a continued succession of fresh air may be obsorbed by the blood.

And in the analysis of the blood, either by fire or fermentation in Exper. 49 and 80, we find good plenty of particles ready to resume the elastick quality of air: But whether any of these air particles enter the blood by the lungs, is not easy to determine; because there is certainly great store of air in the food of animals, whether it be vegetable or animal food. Yet when we consider how much air continually loses its elasticity in the lungs, which seem purposely framed into innumerable minute meanders, that they may thereby the better seize, and bind that volatile *Hermes*: It makes it very probable, that those particles which are now changed from an elastick repulsive, to a strongly attracting state, may easily be attracted thro' the thin partition of the vesicles, by the sulphureous particles which abound in the blood.

And nature seems to make use of the like artifices in

vegetables, where we find that air is freely drawn in; not only with the principal fund of nourishment at the root, but also thro' several parts of the body of the vegetable above ground, which air was seen to ascend in an elastick state most freely and visibly thro' the large *tracheæ* of the Vine; and is thence doubtless carried with the sap into minuter vessels, where being intimately united with the sulphureous, saline and other particles, it forms the nutritive ductile matter, out of which all the parts of vegetables do grow.

Experiment CXI.

It is plain from these effects of the fumes of burning *Brimstone*, lighted Candle, and the breath of Animals on the elasticity in the vesicles of the lungs must be continually decreasing, by reason of the vapours it is there loaded with; so that those vesicles would in a little time subside and fall flat, if they were not frequently replenished with fresh elastick air at every inspiration, thro' which the inferior heated vapour and air ascends, and leaves room for the fresh air to descend into the vesicles, where the heat of the lungs make it expand about $\frac{1}{8}$ part; which degree of expansion of a temperate air, I found by inverting a small glass bubble in water, a little warmer than a *Thermometer* is, by having its ball held some time in the mouth, which may reasonably be taken for the degree of warmth in the cavity of the lungs. When the bubble was cool, the quantity of water imbibed by it was equal to $\frac{1}{8}$ of the cavity of the whole bubble.

But when instead of these frequent recruits of fresh air, there is inspired an air, surcharged with acid fumes and vapours, which not only by their acidity contract the exquisitely sensible vesicles, but also by their grossness much retard the free ingress of the air into the vesicles, many of which are exceeding small, so as not to be visible without a microscope; which fumes are also continually rebating the elasticity of that air; then the air in the vesicles will, by Exper. 107 and 108 lose its elasticity very fast, and consequently the vesicles will fall flat, notwithstanding the endeavours of the extending *Thorax* to dilate them as usual; whereby the motion of the blood thro' the lungs being stopped, instant death ensues.

Which sudden and fatal effect of these noxious vapours, has hitherto been supposed to be wholly owing to the loss and waste of the *vivifying spirit of air*; but may not unreasonably be also attributed to the loss of a considerable part of the air's elasticity, and the grossness and density of the vapours, which the air is charged with; for mutually attracting particles, when floating in so thin a medium as the air, will readily coalesce into grosser combinations: Which effect of these vapours, having not been duly observed before, it was concluded, that they did not affect the air's elasticity; and that consequently the lungs must needs be as much dilated in inspiration by this, as by a clear air.

But that the lungs will not rise and dilate as usual, when they draw in such noxious air, which decreases fast in its elasticity, I was assured by the Experiment I made on my self in Exper. 108, for when towards the latter end of the minute, the suffocating quality of the air in the bladder was greatest, it was with much difficulty that I could dilate my lungs a very little.

From this property in the vapours arising from animal bodies, to rebate and destroy part of the elasticity of the air, a probable account may be given of what becomes of a redundant quantity of air, which may at any time have gotten into the cavity of the *Thorax*; either by a wound, or by some defect in the substance of the lungs, or by very violent exercise. Which if it was to continue always in that expanded state, would very much incommode respiration, by hindering the dilatation of the lungs in inspiration. But if the vapours, which do continually arise in the cavity of the

Thorax, destroy some part of the elasticity of the air, then there will be room for the lungs to heave: And probably, it is in the same manner, that the winds are resorbed, which in their elastick state fly from one part of the body or limbs to another, causing by their distention of the vessels much pain.

Experiment CXII.

I have by the following Experiment found, that the air will pass here and there thro' the substance of the lungs, with a very small force, *viz.*

I cut asunder the bodies of several young and small *animals* just below the *Diaphragm*, and then taking care not to cut any vessel belonging to the lungs, I laid the *Thorax* open by taking away the *Diaphragm*, and so much of the ribs, as was needful to expose the lungs to full view, when blown up. And having cut off the head, I fastned the wind-pipe to a very short inverted leg of a glass syphon; and then placed the inverted lungs and syphon in a large and deep glass vessel *x* full of water (Fig. 32.) under the air pump receiver *p p*, and passing the longer leg of the syphon thro' the top of the receiver, where it was cemented fast at *z*, as I drew the air out of the receiver, the lungs dilated, having a free communication with the outward air, by means of the glass syphon; some of which air would here and there pass in a few places thro' the substance of the lungs, and rise in small streams thro' the water, when the receiver was exhausted no more than to make the *Mercury* in the gage rise less than 2 inches. When I exhausted the receiver, so as to raise the *Mercury* 7 or 8 inches, though it made the air rush with much more violence thro' those small apertures in the surface of the lungs, yet I did not perceive that the number of those apertures was increased, or at least very little. An argument that those apertures were not forcibly made by exhausting the receiver less than two inches, but were originally in the live animal; and that the lungs of living animals are sometimes raised with the like force, especially in violent exercise, I found by the following Experiment, *viz.*

Experiment CXIII.

I tied down a live *Dog* on his back, near the edge of a table, and then made a small hole thro' the intercostal muscles into his *Thorax*, near the *Diaphragm*. I cemented fast into this hole the incurvated end of a glass tube, whose orifice was covered with a little cap full of holes, that the dilatation of the lungs might not at once stop the orifice of the tube. A small vial full of spirit of Wine was tyed to the bottom of the perpendicular tube, by which means the tube and vial could easily yield to the motion of the Dog's body, without danger of breaking the tube, which was 36 inches long. The event was, that in ordinary inspirations, the spirit rose about six inches in the tube; but in great and laborious inspirations, it would rise 24 and 30 inches, *viz.* when I stopped the Dog's nostrils and mouth, so that he could not breathe: This Experiment shews the force with which the lungs are raised by the dilatation of the *Thorax*, either in ordinary or extraordinary and laborious inspirations. When I blew air with some force into the *Thorax*, the Dog was just ready to expire.

By means of another short tube, which had a communication with that which was fixed to the *Thorax* near its insertion into the *Thorax*, I could draw the air out of the *Thorax*, the height of the *Mercury*, instead of spirit in the tube, shewing to what degree the *Thorax* was exhausted of air: The *Mercury* was hereby raised nine inches, which would gradually subside as the air got into the *Thorax* thro' the lungs.

I then laid bare the windpipe, and having cut it off a little below the *Larynx*, I affixed to it a bladder full of air, and then continued sucking air out of the *Thorax*, with a force sufficient to keep the lungs pretty much dilated. As the *Mercury* subsided in the gage, I repeated the suction for a quarter of an hour, till a good part of the air in the bladder was either drawn thro' the substance of the lungs into the *Thorax*, or had lost its elasticity. When I pressed the bladder, the *Mercury* subsided the faster; the Dog was all the while alive, and would probably have lived much longer, if the Experiment had been continued; as is likely from the following Experiment, *viz.*

Experiment CXIV.

I tied a middle-sized Dog down alive on a table, and having layed bare his wind-pipe, I cut it asunder just below the *Larynx*, and fixed fast to it the small end of a common fosset; the other end of the fosset had a large bladder tyed to it, which contained 162 cubick inches; and to the other end of the bladder was tied the great end of another fosset, whose orifice was covered with a valve, which opened inward, so as to admit any air that was blown into the bladder, but none could return that way; yet for further security, that passage was also stopped with spiggot.

As soon as the first fosset was tyed fast to the wind-pipe, the bladder was blown full of air thro' the other fosset; when the Dog had breathed the air in the bladder to and fro for a minute or two, he then breathed very fast, and shewed great uneasiness, as being almost suffocated.

Then with my hand I pressed the bladder hard, so as to drive the air into his lungs with some force; and thereby make his *Abdomen* rise by the pressure of the *Diaphragm*, as in natural breathings: Then taking alternately my hand off the bladder, the lungs with the *Abdomen* subsided; I continued in this manner to make the Dog breathe for an hour; during which time I was obliged to blow fresh air into the bladder every five minutes, three parts in four of that air being either absorbed by the vapours of the lungs, or escaping thro' the ligatures, upon my pressing hard on the bladder.

During this hour, the Dog was frequently near expiring, whenever I pressed the air but weakly into his lungs; as I found by his pulse, which was very plain to be felt in the great crural artery near the groin, which place an assistant held his finger on most part of the time; but the languid pulse was quickly accelerated, so as to beat fast; soon after I dilated the lungs much, by pressing hard upon the bladder, especially when the motion of the lungs was promoted by pressing alternately the *Abdomen* and the bladder, whereby both the contraction and dilatation of the lungs was increased.

And I could by this means rouse the languid pulse whenever I pleased, not only at the end of every 5 minutes, when more air was blown into the bladder from a man's lungs, but also towards the end of the 5 minutes, when the air was fullest of fumes.

At the end of the hour, I intended to try whether I could by the same means have kept the Dog alive some time longer, when the bladder was filled with the fumes of burning *Brimstone*: But being obliged to cease for a little time from pressing the air into his lungs, while matters were preparing for this additional Experiment, in the mean time the Dog dyed, which might otherwise have lived longer, if I had continued to force the air into his lungs.

Now, tho' this Experiment was so frequently disturbed, by being obliged to blow more air into the bladder twelve times during the hour; yet since he was almost suffocated in less than two minutes, by breathing of himself to and fro the first air in the bladder, he would by Experiment 106 on Candles, have dyed in less than two minutes, when one fourth of the old air remained in the bladder, immediately to taint the new admitted air from a man's lungs; so that his continuing to live thro' the whole hour, must be owing to the forcible dilatation of the lungs, by compressing the bladder, and not to the *vivifying spirit of air*. For without that forcible dilatation, he had, after the first 5 or 10 minutes, been certainly dead in less than a minute, when his pulse was so very low and weak, which I did not find to be revived barely by blowing 3 parts in 4 of new air from the lungs of a man into the bladder: But it was constantly roused and quickned, whenever I increased the dilatations of the lungs, by compressing the bladder more vigorously; and that whether it was at the beginning or end of each 5 minutes, yet it was more easily quickned, when the bladder was at any time newly filled, than when it was near empty.

From these violent and fatal effects of very noxious vapours on the respiration and life of animals, we may see how the respiration is proportionably incommoded, when the air is loaded with lesser degrees of vapours, which vapours do in some measure clog and lower the air's elasticity; which it best regains by having these vapours dispelled by the ventilating motion of the free open air, which is rendered wholesome by the agitation of winds: Thus what we call a close warm air, such as has been long confined in a room, without having the vapours in it carried off by communicating with the open air, is apt to give us more or less uneasiness, in proportion to the quantity of vapours which are floating in it. For which reason the *German* stoves, which heat the air in a room without a free admittance of fresh air to carry off the vapours that are raised, as also the modern invention to convey heated air into rooms thro' hot flues, seem not so well contrived, to favour a free respiration, as our common method of fires in open chimneys, which fires are continually carrying a large stream of heated air out of the rooms up the chimney, which stream must necessarily be supplied with equal quantities of fresh air, thro' the doors and windows, or the cranies of them.

And thus many of those who have weak lungs, but can

breathe well enough in the fresh country air, are greatly incommoded in their breathing, when they come into large cities where the air is full of fuliginous vapours, arising from innumerable coal fires, and stenches from filthy lay-stalls and sewers: And even the most robust and healthy in changing from a city to a country air, find an exhilarating pleasure, arising from a more free and kindly inspiration, whereby the lungs being less loaded with condensing air and vapours, and thereby the vesicles more dilated, with a clearer and more elastick air, a freer course is thereby given to the blood, and probably a purer air mixed with it; and this is one reason why in the country a serene dry constitution of the air is more exhilarating than a moist thick air.

And for the same reason, 'tis no wonder, that pestilential, and other noxious epidemical infections are conveyed by the breath to the blood (when we consider what great quantities of the airy vehicle loses its elasticity among the vesicles, whereby the infectious *Miasma* is lodged in the lungs.

When I reflect on the great quantities of elastick air, which are destroyed by burning sulphur; it seems to me not improbable, that when an animal is killed by lightning without any visible wound, or immediate stroke, that it may be done by the air's elasticity, being instantly destroyed by the sulphureous lightning near the animal, whereby the lungs will fall flat, and cause sudden death; which is further confirmed by the flatness of the lungs of animals thus killed by lightning, their vesicles being found upon dissection to be fallen flat, and to have no air in them: The bursting also of glass windows outwards, seems to be from the same effect of lightning on the air's elasticity.

It is likewise by destroying the air's elasticity in fermented liquors, that lightning renders them flat and vapid: For since sulphureous steams held near or under vessels will check redundant fermentation, as well as the putting of sulphureous mixtures into the liquor, 'tis plain, those steams can easily penetrate the wood of the containing vessels. No wonder then, that the more subtile lightning should have the like effects. I know not whether the common practice of laying a bar of iron on a vessel, be a good preservative against the ill effects of lightning on liquors. I should think that the covering a vessel with a large cloth dipped in a strong brine, would be a better preservative; for salts are known to be strong attracters of sulphur.

The certain death which comes on the explosion of Mines, seems to be effected in the same manner: For tho' at first there is a great expansion of the air, which must dilate the lungs, yet that air is no sooner filled with fuliginous vapours, but a good deal of its elasticity is immediately destroy'd: As in the case of burning Matches in Experiment 103, the heat of the flame at first expanded the air; but notwithstanding the flame continued burning, it immediately contracted, and lost much of its elasticity, as soon as some quantity of sulphureous steams ascended in it.

Which steams have doubtless the same effect on the air, in the lungs of animals held over them; as in the *Grotto di cani*, or when a close room is filled with them, where they certainly suffocate.

It is found by Experiments 103, 106, and 107, that an air greatly charged with vapours loses much of its elasticity, which is the reason why subterraneous damps suffocate Animals, and extinguish the flame of Candles. And by Experiment 106, we see that the sooner a Candle goes out, the faster the air loses its elasticity.

Experiment CXV.

This put me upon attempting to find some means to qualify and rebate the deadly noxious quality of these vapours: And in order to it, I put thro' the hole, in the top of the air pump receiver (Fig. 32.) which contained two quarts, one leg of an iron syphon made of a gun-barrel, which reached near to the bottom of the receiver: It was cemented fast at *z*, I tyed three folds of woollen cloth over the orifice of the syphon, which was in the receiver. The Candle went out in less than two minutes, tho' I continued pumping all the while, and the air passed so freely thro' the folds of cloth into the receiver, that the *Mercury* in the gage did not rise above an inch.

When I put the other end of the syphon into a hot iron pot, with burning *Brimstone* in it; upon pumping, the Candle went out in 15 seconds of a minute; but when I took away the 3 folds of cloth, and drew the sulphureous steams thro' the open syphon, the light of the Candle was instantly extinguished; whence we see the 3 folds of cloth preserved the Candle alight 15″. And where the deadly quality of vapours in Mines is not so strong as these sulphureous ones were, the drawing the breath thro' many folds of woollen cloth may be a means to preserve life a little longer, in proportion to the more or less noxious quality of the damps.

When, instead of the 3 folds of cloth, I immersed the end of the syphon 3 inches deep in water in the vessel *x*, (Fig. 32.) tho' upon pumping the sulphureous fumes did ascend visibly through the water, yet the Candle continued burning half a minute, *i. e.* double the time that it did when fumes passed thro' folds of woollen cloth.

Experiment CXVI.

I bored a hole in the side of a large wooden fosset *a b*, (Fig. 39.) and glewed into it the great end of another fosset *i i*, covering the orifice with a bladder valve *r*: Then I fitted a valve *b i*, to the orifice of the iron syphon *s s*, fixing the end of the syphon fast at *b* into the fosset *a b*: Then by means of narrow hoops I placed four *Diaphragms* of flannel at half an inch distance from each other, into the broad rim of a sieve, which was about 7 inches diameter. The sieve was fixed to, and had a free communication with both orifices of the syphon, by means of two large bladders *i i n n o*.

The instrument being thus prepared, pinching my nostrils close, when I drew in breath with my mouth at *a*, the valve *i b* being thereby lifted up, the air passed freely through the syphon from the bladders, which then subsided, and shrunk considerably: But when I breathed air out of my lungs, then the valve *i b* closing the orifice of the syphon, the air passed thro' the valve *r* into the bladders, and thereby dilated them; by which artifice the air which I expired must necessarily pass thro' all the *Diaphragms*, before it could be inspired into my lungs again. The whole capacity of the bladders and syphon was 4 or 5 quarts.

Common sea salt, and *Sal Tartar*, being strong imbibers of sulphureous steams, I dipped the four *Diaphragms* in strong solutions of those salts, as also in white wine vinegar, which is looked upon as a good anti-pestilential: Taking care after each of these Experiments to cleanse the syphon and bladder well from the foul air, by filling them with water.

I could breathe too and fro the air inclosed in this instrument for a minute and half, when there were no *Diaphragms* in it; when the 4 *Diaphragms* were dipped in vinegar, 3 minutes; when dipped in a strong solution of sea-salt, 3 minutes and an half. In a Lixivium of *Sal Tartar*, 3 minutes; when the *Diaphragms* were dipped in the like Lixivium, and then well dryed, 5 minutes; and once $8+\frac{1}{2}$ minutes, with very highly calcined *Sal Tartar*; but whether this was owing to the *Tartar*'s being greatly calcined, whereby it might more strongly attract sulphureous gross vapours, or whether it was owing to the bladder and syphon's being entirely dry, or whether it was occasioned by some unheeded passage for the air thro' the ligatures, I am uncertain; neither did I care to ascertain the matter by repeated Experiments, fearing I might thereby some way injure my lungs, by frequently breathing in such gross vapours.

Hence *Sal Tartar* should be the best preservative against noxious vapours, as being a very strong imbiber of sulphureous, acid and watry vapours, as is sea salt also: For having carefully weighed the 4 *Diaphragms*, before I fix't them in the instrument, I found that they had increased in weight 30 grains in five minutes; and it was the same in two different tryals; so they increased in weight at the rate of 19 ounces in 24 hours. From which deducting $\frac{1}{6}$ part for the quantity of moisture, which I found those *Diaphragms* attracted in 5 minutes in the open air; there remains $15+\frac{2}{3}$ ounces, for the weight of the moisture from the breath in 24 hours: But this is probably too great an allowance, considering that the *Diaphragms* might attract more than $\frac{1}{6}$ part from the moisture of the bladders and of the syphon.

I have found that when the *Diaphragms* had some small degree of dampness, they increased in weight six grains in 3 minutes; but they made no increase in weight in the same time, when in the open air: which six grains in 3 minutes, is at the rate of about $6+\frac{1}{2}$ ounces in 24 hours; and this is nearly the same proportion of moisture that I obtained by breathing into a large receiver full of spunges. But the 6 grains imbibed by the four *Diaphragms* in 3 minutes, was not near all the vapours which were in that bulk of inclosed air; for at the end of the 3 minutes, the often respired air was so loaded with vapours, which in that floating state were easily, by their mutual attraction, formed into combinations of particles, too gross to enter the minute vesicles of the lungs, and was therefore unfit for respiration; so that it is not easie to determine what proportion is carried off by respiration, especially considering that some of the inspired air, which has lost its elasticity in the lungs, is mingled with it. But supposing $6+\frac{1}{2}$ ounces to be the quantity of moisture carried off by respiration in 24 hours, then the surface of the lungs being found as above 41635 square inches only $\frac{1}{3751}$ part of an inch depth, will be evaporated off their inward surface in that time, which is but $\frac{1}{75}$ part of the depth of what is perspired off the surface of a man's body in that time.

If then life can by this means be supported for 5 minutes with 4 *Diaphragms* and a gallon of air, then doubtless, with double that quantity of air and 8 *Diaphragms* we might well expect to live at least 10 minutes. It was a considerable disadvantage that I was obliged to make use of bladders, which had been often wetted and dried, so that the unsavoury fumes from them must needs have contributed much to the unfitting the included air for respiration: Yet there is a necessity for making use of either bladder or leather in these cases; for we cannot breathe to and fro the air of a vessel, whose sides will not dilate and contract in conformity with the expirations and inspirations, unless the vessel be very large, and too big to be conveniently portable.

Having stopped up the wide sucking orifice of a large pair of kitchen bellows, they being first dilated, I could breathe to and fro at their nose, the air contained in them for 3 minutes, without much inconvenience, they heaving and falling very easily by the action of respiration. Some such like instrument might be of use in any case where a room was filled with suffocating vapours, where it might be necessary to enter for a few minutes, in order to remove the cause of them, or to fetch any person or thing out; as in the case when houses are first beginning to fire, in the chymists elaboratories; and in many other cases where places were filled with noxious deadly vapours, as in the case of stink pots thrown into ships, in mines, *&c.*

But in every *apparatus* of this kind great care must always be taken, that the inspiration be as free as possible, by making large passages and valves to play most easily. For tho' a man by a peculiar action of his mouth and tongue may suck *Mercury* 22 inches, and some men 27 or 28 high; yet I have found by experience, that by the bare inspiring action of the *Diaphragm*, and dilating *Thorax*, I could scarcely raise the *Mercury* 2 inches. At which time the *Diaphragm* must act with a force equal to the weight of a *Cylinder* of *Mercury*, whose base is commensurate to the area of the *Diaphragm*, and its height 2 inches, whereby the *Diaphragm* must at that time sustain a weight equal to many pounds. Neither are its counter-acting muscles, those of the *Abdomen*, able to exert a greater force.

For notwithstanding a man, by strongly compressing a quantity of air included in his mouth, may raise a column of *Mercury* in an inverted syphon, to 5 or 7 inches height, yet he cannot with his utmost strainings raise it above 2 inches, by the contracting force of the muscles of the *Abdomen*; whence we see that our loudest vociferations are made with a force of air no greater than this. So that any small impediment in breathing will hasten the suffocation, which consists chiefly in the falling flat of the lungs, occasioned by the grossness of the particles of a thick noxious air, they being in that floating state most easily attracted by each other: As we find in the foregoing experiments that sulphur and the elastick repelling particles of air do: And subsequently unelastick, sulphureous, saline and other floating particles will most easily coalesce, whereby they are rendred too gross to enter the minute vesicles; which are also much contracted, as well by the loss of the elasticity of the contained air, as by the contraction occasioned by the stimulating, acid, sulphureous vapours. And 'tis not improbable that one great design of nature, in the structure of this important and wonderful *viscus*, was to frame its vesicles so very minute, thereby effectually to hinder the ingress of gross feculent particles, which might be injurious to the animal œconomy.

This quality of salts strongly to attract sulphureous, acid and other noxious particles, might make them very beneficial to mankind in many other respects. Thus in several unwholesome trades, as the smelters of metals, the ceruss-makers, the plumbers, *&c.* it might not unlikely be of good service to them in preserving them in some measure at least, from the noxious fumes of the materials they deal in, which by many of the foregoing experiments we are assured must needs coalesce with the elastick air in the lungs, and be lodged there; to prevent which inconvenience the workmen might, while they are at work, make use of pretty broad mufflers, filled with 2, 4, or more *Diaphragms* of flannel or cloth dipped in a solution of *Sal Tartar* or *Pot-ash*, or Sea Salt and then dryed.

The like mufflers might also be of service in many cases where persons may have urgent occasion to go for a short time into an infectious air: Which mufflers might, by an easy contrivance, be so made as to draw in breath thro' the *Diaphragms*, and to breathe it out by another vent.

In these and the like cases this kind of mufflers may be very serviceable; but in the case of the damps of mines they are by no means to be depended on, because they are not a sufficient screen from so very noxious vapours.

Experiment CXVII.

We have from the following Experiment a good hint, to make these Salts of service to us in some other respects, &c.

I set a lighted *Candle* under a large receiver (Fig. 35.) which contained about 4 gallons, it continued burning for $3+\frac{1}{2}$ minutes, in which time it had absorbed about a quart of air. I then filled the receiver with fresh air, by pouring it full of water, and then emptying of it; when having wiped it dry, I lined all the inside with a piece of flannel dipped in a lixivium of *Sal Tartar*, and then dryed; the flannel was extended with little hoops made of pliant twigs. The *Candle* continued burning under the receiver thus prepared $3+\frac{1}{2}$ minutes, yet it absorbed but two thirds of the quantity of air which it absorbed when there was no flannel in the receiver.

The reason of which difference in the quantities of elastick air absorbed, appears from Experiment 106. where least air was always absorbed in least receivers, which was the present case: For the flannel lining, besides the space it took up, could not be so closely adapted, but that there was left a full third of the capacity of the receiver, between the lining and the receiver: So that the *Candle* burnt in a bulk of air less by one third than the whole capacity of the receiver; for which reason less air also was absorbed.

And we may further observe, that since the *Candle* continued burning as long in a quantity of air, equal but to two thirds of the receiver, as in the whole air of the receiver; this must be owing to the *Sal Tartar* in the flannel lining, which must needs have absorbed one third of the fuliginous vapours, which arose from the burning Candle. Hence we may not unreasonably conclude, that the pernicious quality of noxious vapours in the air might, in many cases, be much rebated and qualified by the strongly absorbing power of Salts.

Whether Salts will have a good effect in all, or any of these cases, experience will best inform us. There is certainly sufficient ground, from many of the foregoing Experiments, to encourage us to make the tryal, and they may at least be hints for further improvements.

We see that Candles and burning *Brimstone* do in a much greater degree destroy the elasticity of the air, than the breath of Animals; because their vapours are more plentiful, and abound more with acid sulphureous particles, and are also less diluted with watry vapours, than the breath of Animals is: In which also there are sulphureous particles, tho' in lesser degrees; for the animal fluids, as well as solids, are stored with them: And therefore the Candle and Matches ceasing to burn, soon after they are confined in a small quantity of air, seems not to be owing to their having rendred that air effete, by having consumed its *vivifying spirit*; but should rather be owing to the great quantity of acid fuliginous vapours, with which that air is charged, which destroy a good deal of its elasticity, and very much clog and retard the elastick motion of the remainder.

And the effect the half exhausting of a receiver has upon the elasticity of the remaining half of the air, seems to be the reason why the flame of a Candle does not continue burning, till it has filled the receiver it stands in with fumes, but goes out the quicker, the sooner the air is drawn out to that degree; which seems therefore to be owing to this, that an air rarified to double its space, will not expand so briskly with the warmth of flame, as a more condensed air will do: And consequently action and re-action being reciprocal, will not give so brisk a motion to the flame, which subsists by a constant succession of fresh air, to supply the place of the either absorbed, or much dilated air, which is continually flying off. And the quicker the succession of this fresh air is, by blowing, the more vigorously does a fire burn.

If the continuance of the burning Candle be wholly owing to the *vivifying spirit*, then supposing in the case of a receiver, capacious enough for a Candle to burn a minute in it, that half the *vivifying spirit* be drawn out with half the air, in ten seconds of time; then the Candle should not go out at the end of those 10 seconds, but burn 20 seconds more, which it does not; therefore the burning of the Candle is not wholly owing to the *vivifying spirit*, but to certain degrees of the air's elasticity. When a wholly exhausted receiver was by means of a burning glass first filled with the fumes of brown paper with *Nitre*, and then filled with fresh air, the nitrous paper upon applying the burning glass did freely detonize; and a Candle put into a like air, burnt for 28″; which in a fresh air, in the same receiver, burnt but 43″; but when the same receiver with air in it, was filled full of fumes of detonized *Nitre*, and a Candle placed in that thick vapour, it went out instantly, for a Candle will not burn, nor the *Nitre* detonize in a very rare, nor a very thick air; whence the reason why the *Nitre* detonized, and the Candle burnt, when placed in the receiver, after fresh air was let in upon the fumes which were made *in vacuo*, was that those fumes were much dispersed and condensed on the sides of the glass, upon the rushing in of the fresh air, for the fumes were then much more rare and transparent, than before the air was let in.

That a Fire which is supplied with a hot air will not burn so briskly as a Fire which is fed by a cool air is evident from hence; that when the *Sun* shines on a Fire, and thereby too much rarifies the ambient air, that Fire will not burn well, nor will a small Fire burn so well near a large one as at some distance from it. And *e contra*, it is a common observation, that in very cold frosty weather, Fires burn most briskly; the reason of which seems to be this, that the elastick expansion of the cold condensed air to a rarified state, when it enters the Fire, is much brisker than that of an air already rarified in a good measure by heat, before it enters the Fire; and consequently a continued succession of cold air must give a brisker motion to the Fire, than the like succession of hot air: And such colder and more condensed air will also (as Sir *Isaac Newton* observes, qu. 11.) by its greater weight check the ascent of the vapours and exhalations of the Fire, more than a warmer lighter air. So that between the action and re-action of the air and sulphur of the fuel, and of the colder and denser circumambient air, which rarifies much upon entering the Fire, the heat of the Fire is greatly increased.

This continual supply of fresh air to the fuel seems hence also very necessary for keeping a Fire alive; because it is found, that a *Brimstone Match* will not take Fire in *a vacuum*, but only boil and smoak; nor will *Nitre* incorporated into *Brown Paper* then detonize, except here and there a single grain, that part only of the *Paper* turning black on which the focus of the burning glass falls; nor would they burn when a half exhausted receiver with fumes in it was filled with fresh air added to those fumes: In which case it is plain, that a good quantity of the supposed *vivifying spirit of air* must enter the receiver with the fresh air, and consequently those substances should take fire, and burn for a short time at least, which yet they did not.

And that the air's elasticity conduces much to the intense burning of Fires, seems evident from hence; that *Spirit* of *Nitre* (which by Experiment 75 has but little elastick air in it) when poured upon live *Coals*, extinguishes instead of invigorating them: But *Spirit* of *Nitre*, when by being mixt with *Sal Tartar* it is reduced to *Nitre*, will then flame, when thrown into the Fire, *viz.* because *Sal Tartar* abounds with elastick aereal particles, as appears by Experiment 74, where 224 times its bulk of air arose from a quantity of *Sal Tartar*. And for the same reason it is that common *Nitre*, when thrown into the Fire, flames, tho' its *Spirit* will not, *viz.* because there is much elastick air in it, as appears from Experiment 72, as well as from the great quantity of it, generated in the firing of *Gun-powder*.

The reason why *Sal Tartar*, when thrown on live Coals, does not detonize and flame like *Nitre*, (notwithstanding by Experiment 74 plenty of elastick particles did arise from it) is this, *viz.* because by the same Experiment, compared with Experiment 72, it is found, that a much more intense degree of heat was required to extricate the elastic air from *Sal Tartar*, the more fix'd body, than from *Nitre;* the great degree of Fire with which *Sal Tartar* is made, rendering the cohesion of its parts more firm: For it is well known that fire, instead of disuniting, does in many cases inseparably unite the parts of bodies: And hence it is that *Pulvis Fulminans*, which is a mixture of *Sal Tartar, Nitre* and sulphur, gives a greater explosion than *Gunpowder*: Because the particles of the *Sal Tartar*, cohering more firmly in a fix'd state than those of *Nitre*, they are therefore thrown off with a great repulsive force, by the united action and reaction of all those ingredients armed each with its acid *Spirit*.

Experiment CXVIII.

Which acid *Spirits* consisting of a volatile acid *Salt* diluted in phlegm do contribute much to the force of explosion; for when heated to a certain degree, they make a great explosion, like water heated to the same degree, as I found by dropping a few drops of *Spirit* of *Nitre*, oil of *Vitriol*, water, and spittle on an *Anvil*; and then holding over those drops a piece of *Iron* which had a white heat given it; upon striking down the hot *Iron* with a large Hammer, there was a very great explosion made by each of those liquors: But frothy spittle, which had air in it, made a louder explosion than water; which shews that the vast explosion of the *Nitre* and *Sal Tartar*, which are composed of elastick air particles, included in an acid *Spirit*, is owing to their united force.

We may therefore from what has been said, with good reason conclude, that Fire is chiefly invigorated by the action and re-action of the acid sulphureous particles of the fuel, and the elastick ones which arise and enter the Fire, either from the fuel in which they abound, or from the circumambient air: For by Experiment 103, and many others, acid sulphureous particles act vigorously on air; and since action and re-action are reciprocal, so must air on sulphur; and there is we see, plenty of both, as well in mineral as vegetable fuel, as also in animal substances, for which reason they will burn.

But when the acid sulphur, which we see acts vigorously on air, is taken out of any fuel, the remaining *Salt*, Water and Earth are not inflammable, but on the contrary quench and retard fire; and as air cannot produce fire without sulphur, so neither can sulphur burn without air: Thus *Charcoal* heated to an intense degree for many hours in a close vessel will not burn as in the open air, it will only be red hot all the time like a mass of Gold without wasting: But no sooner is it exposed to the free air, but the sulphur, by the violent action and re-action between that and the elastick air, is soon separated and carried off from the *Salt* and Earth, which are thereby reduced from a solid and hard to a soft impalpable *calx*.

And when a *Brimstone Match* which was placed in an exhausted receiver was heated by the *focus* of a burning glass so as to melt the *Brimstone*, yet it did not kindle into fire nor consume, notwithstanding the strength and vigour of the action and re-action that is observed between light and sulphureous bodies. Which is assigned by the illustrious Sir *Isaac Newton*, as "one reason why sulphureous bodies take fire more readily, and burn more vehemently than other bodies do, qu. 7." What his notion of fire and flame is, he gives us in qu. 9. and 10. Qu. 9. "Is not fire a body heated so hot as to emit light copiously? For what else is a red hot *Iron* than fire? And what else is a burning *Coal*, than red hot *Wood*? Qu. 10. Is not flame a vapour, fume or exhalation heated red hot, that is so hot as to flame? For bodies do not flame without emitting a copious fume, and this fume burns in the flame.——Some bodies heated by motion or fermentation, if the heat grow intense, fume copiously, and if the heat be great enough, the fumes will shine and become flame: Metals in fusion do not flame for want of a copious fume, except spelter which fumes copiously, and thereby flames: All flaming bodies, as Oil, Tallow, Wax, Wood, fossil Coals, Pitch, Sulphur, by flaming waste and vanish into burning smoak; which smoak, if the flame be put out, is very thick and visible, and sometimes smells strongly, but in flame loses its smell by burning; and according to the nature of the smoak the flame is of several colours, as that of sulphur, blue; that of copper opened with sublimate, green; that of tallow, yellow; that of camphire, white; smoak passing thro' flame cannot but grow red hot, and red hot smoak can have no other appearance than that of flame."

But Mr. *Lemery* the younger says, "that the matter of light produces sulphur, being mixt with compositions of

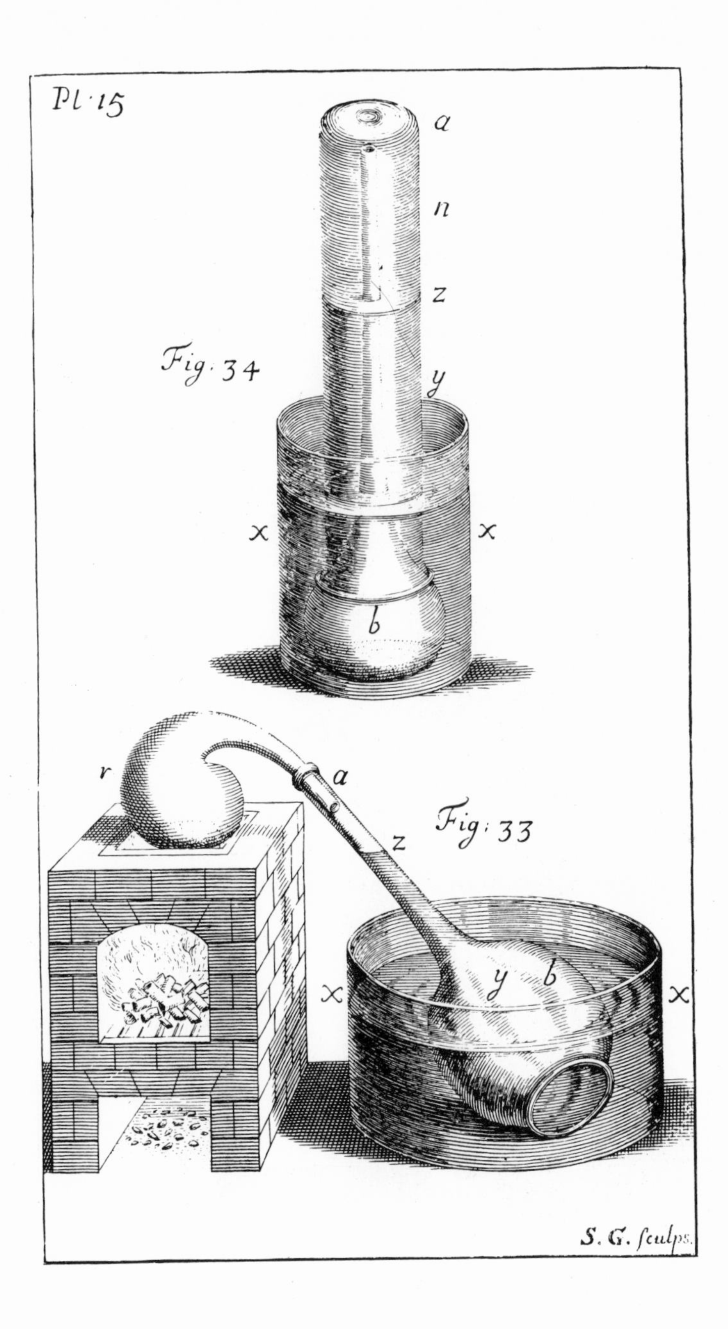
Pl. 15
a
n
z
Fig. 34
y
x
x
b
r
a
Fig. 33
z
x
y
b
x
S. G. sculps.

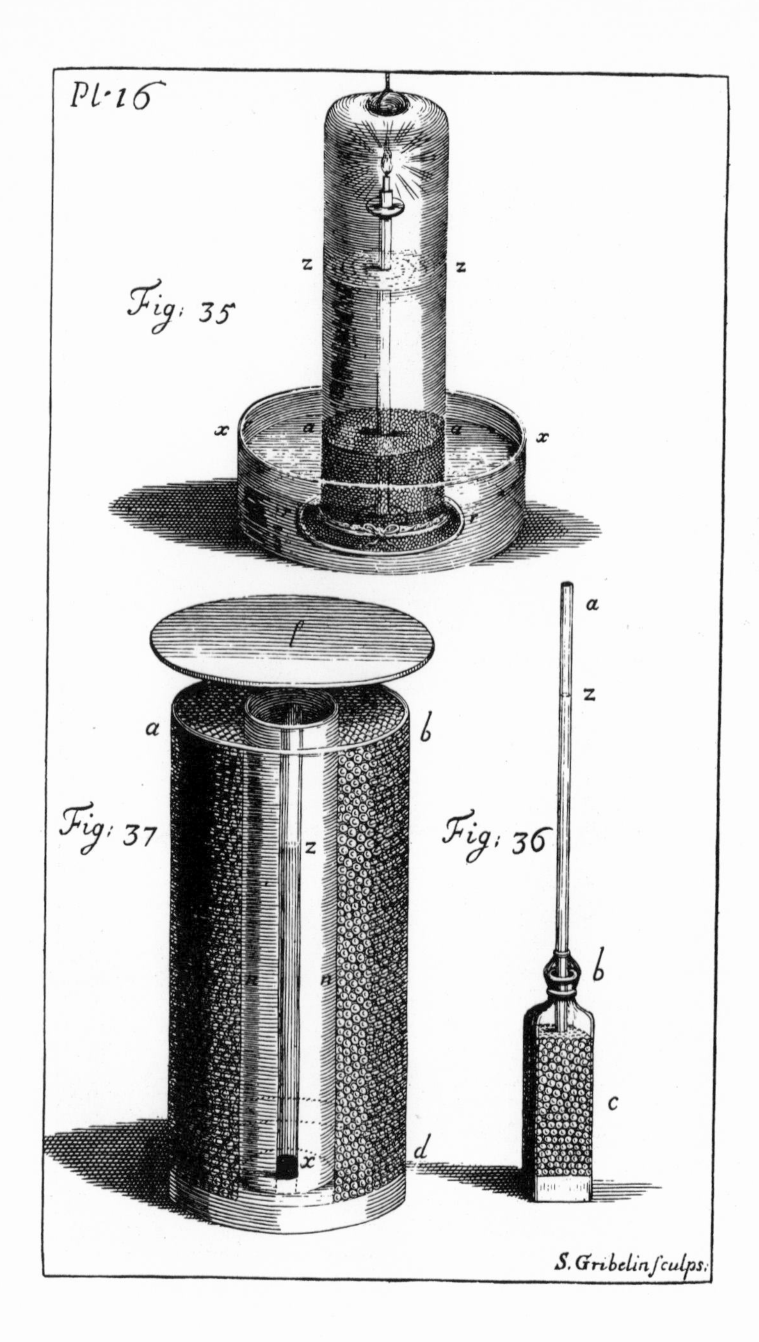
Pl. 16
Fig: 35
z
z
x
a
a
x
Fig: 37
f
a
b
z
x
d
Fig: 36
a
z
b
c
S. Gribelin sculps.

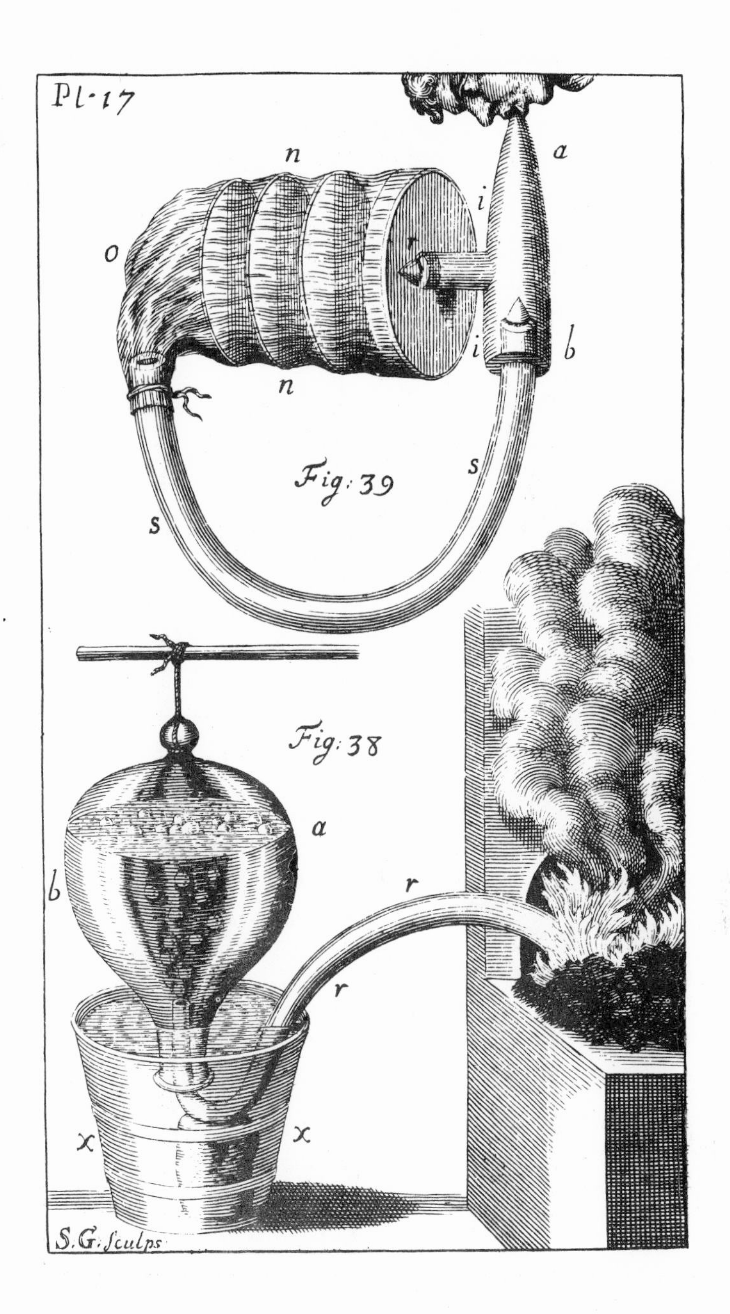
Pl·17
n
a
i
o
r
i
b
n
s
Fig: 39
s
Fig: 38
a
b
r
r
x
x
S.G. sculps

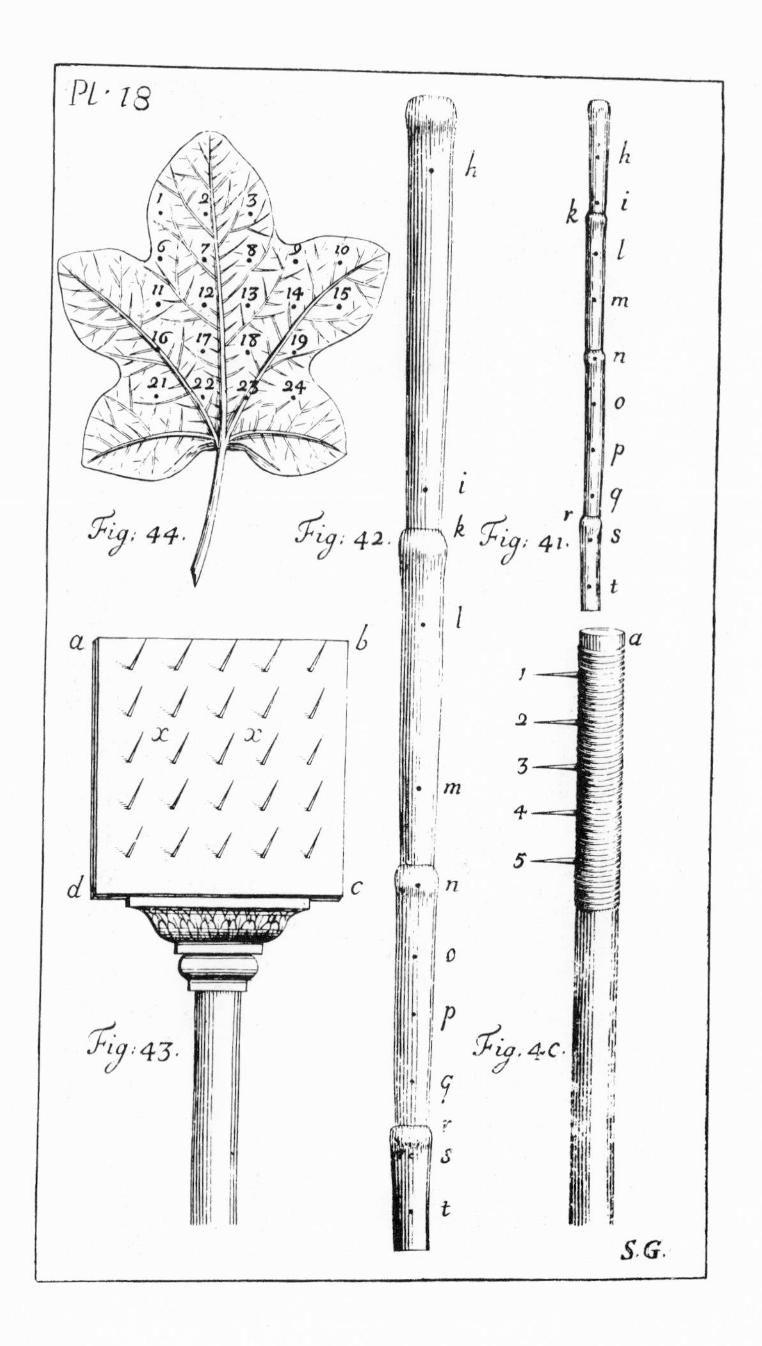
Pl. 18
1
2
3
6
7
8
9
10
11
12
13
14
15
16
17
18
19
21
22
23
24
Fig: 44.
Fig: 42.
Fig: 41.
h
i
k
l
m
n
o
p
q
r
s
t
a
b
c
d
x
x
Fig: 43.
Fig. 40.
1
2
3
4
5
S.G.

salt, earth and water, and that all inflammable matters are such only in vertue of the particles of fire which they contain. For in the Analysis, such inflammable bodies produce salt, earth, water, and a certain subtle matter, which passes thro' the closest vessels, so that what pains soever the artist uses, not to lose any thing, he still finds a considerable diminution of weight.

"Now these principles of salt, earth and water are inactive bodies, and of no use, in the composition of inflammable bodies, but to detain and arrest the particles of fire, which are the real and only matter of flame.

"It appears therefore to be the matter of flame that the artist loses in decompounding inflammable bodies, *Mem. de l' Acad. Anno* 1713."

But by many of the preceding Experiments, it is evident, that the matter lost in the Analysis of these bodies was elastick air, a very active principle in fire, but not an elemental fire, as he supposes.

"Mr. *Geoffrey* compounded sulphur of acid Salt, Bitumen, a little Earth and *oil* of *Tartar*." *Mem. de l' Acad. Anno* 1703. In which *oil* of *Tartar* there is much air by Experiment 74, which air was doubtless by its elasticity very instrumental in the inflammability of this artificial sulphur.

If fire was a particular distinct kind of body inherent in sulphur, as Mr. *Homberg*, Mr. *Lemery*, and some others imagin, then such sulphureous bodies, when ignited, should rarify and dilate all the circumambient air; whereas it is found by many of the preceding Experiments, that acid sulphureous fuel constantly attracts and condenses a considerable part of the circumambient elastick air. An argument that there is no fire endued with peculiar properties inherent in sulphur; and also that the heat of fire consists principally in the brisk vibrating action and re-action, between the elastick repelling air, and the strongly attracting acid sulphur, which sulphur in its Analysis is found to contain an inflammable oil, and acid salt, a very fix't earth, and a little metal.

Now sulphur and air are supposed to be acted by that ethereal medium, "by which (the great Sir *Isaac Newton* supposes) light is refracted and reflected, and by whose vibrations light communicates heat to bodies, and is put into fits of easie reflection, and easie transmission: And do not the vibrations of this medium in hot bodies contribute to the intenseness and duration of their heat? And do not hot bodies communicate their heat to contiguous cold ones, by the vibrations of this medium, propagated from them into cold ones? And is not this medium exceedingly more rare and subtle than the air, and exceedingly more elastick and active? And does it not readily pervade all bodies, *Optick qu.* 18. The elastick force of this medium, in proportion to its density must be above 490,000,000,000 times greater than the elastick force of the air is, in proportion to its density, *ibid. qu.* 21." A force sufficient to give an intense degree of heat, especially when its elasticity is much increased by the brisk action and re-action of particles of the fuel and ambient air.

From this manifest attraction, action and re-action, that there is between the acid, sulphureous and elastick aereal particles, we may not unreasonably conclude, that what we call the fire particles in Lime, and several other bodies, which have undergone the fire, are the sulphureous and elastick particles of the fire fix't in the Lime; which particles, while the Lime was hot, were in a very active, attracting and repelling state; and being, as the Lime cooled, detained in the solid body of the Lime, at the several attracting and repelling distances, they then happened to be at, they must necessarily continue in that fix't state, notwithstanding the ethereal medium, which is supposed freely to pervade all bodies, be continually solliciting them to action: But when the solid substance of the Lime is dissolved, by the affusion of some liquid, being thereby emancipated, they are again at liberty to be influenced and agitated by each other's attraction and repulsion, upon which a violent ebullition ensues, from the action and re-action of these particles, which ebullition ceases not, till one part of the elastick particles are subdued and fix't by the strong attraction of the sulphur, and the other part is got beyond the sphere of its attraction, and thereby thrown off into true permanent air: And that this is a probable solution of the matter, there is good reason to conclude, from the frequent instances we have in many of the foregoing Experiments, that plenty of elastick air is at the same time both generated and absorbed by the same fermenting mixture; some of which were observed to generate more air than they absorbed, and others *e contra* absorbed more than they generated, which was the case of Lime.

Experiment CXIX.

And that the sulphureous and aereal particles of the fire are lodged in many of those bodies which it acts upon, and thereby considerably augments their weight, is very evident in Minium or Red Lead, which is observed to increase in weight about $\frac{1}{20}$ part in undergoing the action of the fire. The acquired redness of the Minium, indicating the addition of plenty of sulphur in the operation: For sulphur, as it is found to act most vigorously on light, so it is apt to reflect the strongest, *viz.* the red rays; and that there is good store of air added to the Minium, I found by distilling first 1922 grains of Lead, from whence I obtained only seven cubick inches of air; but from 1922 grains, which was a cubick inch of Red Lead, there arose in the like space of time 34 cubick inches of air; a great part of which air was doubtless absorbed by the sulphureous particles of the fuel, in the reverberatory furnace, in which the Minium was made; for by Experiment 106. the more the fumes of a fire are confined, the greater quantity of elastick air they absorb.

It was therefore doubtless this quantity of air in the Minium which burst the hermetically sealed glasses of the excellent Mr. *Boyle*, when he heated the Minium contained in them by a burning glass; but the pious and learned Dr. *Nieuwentyt* attributes this effect wholly to the expansion of the fire particles lodged in the Minium, "he supposing fire to be a particular fluid matter, which maintains its own essence, and figure, remaining always fire, tho' not always burning. *Religious Philosopher, p.* 310."

To the same cause also, exclusive of the air, he attributes the vast expansion of a mixture of *compound Aqua fortis* and *oil* of *Carraways,* whereas by Exper. 62. there is a great quantity of air in all *oils*. And by pouring some *compound Aqua-fortis* on *oil* of *Cloves*, the mixture expanded into a space equal to 720 times the bulk of the *oil*, that part of the expansion, which was owing to the watry part of the *oil* and *spirit* was soon contracted; whereas the other part of the expansion, which was owing to the elastick air of the *oil*, was not all contracted, till the next day, by which time the sulphureous fumes had resorbed it.

The learned *Boerhaave* would have it, that putrefaction is the effect of inherent fire. He says, "that vegetables alone are the subject of fermentation, but both vegetables and animals of putrefaction; which operations he attributes to very different causes, the immediate cause of fermentation is (he says) the motion of the air intercepted between the fluid and viscous parts of the fermenting liquor; but the cause of putrefaction is fire itself, collected or included within the putrefying subject, *Process*. 77." But I do not see why these may not reasonably enough be looked upon as the effects of different degrees of fermentation; nutrition being the genuine effect of that degree of it, in which the sum of the attracting action of the particles is much superior to the sum of their repulsive power: But when their repelling force far exceeds their attractive, then the component parts of vegetables are dissolved. Which dissolving substances, when they are diluted with much liquor, do not acquire a great heat in the dissolution, the briskness of the intestine motion being checked by the liquor: But when they are only moist, like green and damp Hay, in a large heap, then they acquire a violent heat, so as to scorch, burn and flame, whereby the union of their constituent parts being more throughly dissolved, they will neither produce a vinous, nor an acid spirit: Which great degree of solution may well be effected by this means, without the action of a fire, supposed to be included within the putrefying subject. Wherefore according to the old Axiom, *Entia non sunt temere neque absque necessitate multiplicanda.*

If the notion of fermentation be restrained to the greater repelling degrees of fermentation, in which sense it has commonly been understood; then it is as certain, that the juices of vegetables and animals do not ferment in a healthy state, as it is, that they do not at the same time coalesce and disunite: But if fermentation be taken in a larger sense, for any the smallest to the greatest degree of intestine motion of the particles of a fluid, then all vegetable and animal fluids are in a natural state, in some degree of ferment, for they abound both with elastick and sulphureous particles: And it may with as much reason be argued, that there is no degree of warmth in animals and vegetables, because a great degree of heat will cause a solution of continuity, as to say, there is no degree of ferment in the fluids of those bodies, because a great repelling degree of ferment will most certainly dissolve them.

That illustrious Philosopher Sir *Isaac Newton*, in his thoughts about the nature of acids, gives this rational account of the nature of fermentation. "The particles of acids—are endued with a great attractive force, in which force their activity consists—By this attractive force they get about the particles of bodies, whether they be of a metallick or stony nature, and adhere to them most closely on all sides, so that they can scarce be separated from them, by distillation or sublimation; when they are attracted and gathered together about the particles of bodies, they raise, disjoyn, and shake them one from another, that is, they dissolve those bodies.

"By their attractive force also, by which they rush towards the particles of bodies, they move the fluid, and excite heat, and they shake asunder some particles, so much as to turn them into air, and generate bubbles: And this is the reason of dissolution, and all violent fermentation. *Harris Lexicon Tech.* Vol. II. introduction."

Thus we have from these Experiments many manifest proofs of considerable quantities of true permanent air, which are by means of fire and fermentation raised from, and absorbed by animal, vegetable and mineral substances.

That this air consists of particles which are in a very active state, repelling each other with force, and thereby constituting the same kind of elastick fluid with common air, is plain from its raising the *Mercury* in Experiment 88 and 89, and from its continuing in that elastick state for many months, tho' cooled by severe frosts; whereas watry vapours, tho' they expand much with heat, yet are found immediately to condense into their first dimensions when cold.

The air generated by fire was not, in many instances, separated without great violence from the fix'd bodies, in which it was incorporated; as in the case of *Nitre, Tartar, Sal Tartar* and *Copperas*: whence it should seem, that the air generated from these Salts, may probably be very instrumental in the union of Salts, as well as that central, denser, and compacter particle of earth, which Sir *Isaac Newton* observes, does by its attraction make the watry acid flow round it, for composing the particles of Salt. *qu.* 31. For since upon the dissolution of the constituent parts of Salt by fire, it is found, that upon separating and volatilizing the acid spirit, the air particles do in great abundance rush forth from a fixt to a repelling elastick state; it must needs be, that these particles did in their fixt state strongly attract the acid spirits, as well as the sulphureous earthy parts of the Salt; for the most strongly repelling and elastick particles are observed, in a fixt state, to be the most strongly attracting.

But the watry acid, which when separated from Salt by the action of fire, makes a very corrosive fuming spirit, will not make elastick air, tho' its parts were put into a brisk motion by fire in Exper. 75. And the event was the same with several other volatile substances, as volatile Salt of *Sal Ammoniac, Camphire* and *Brandy*, which tho' distilled over with a considerable heat, yet generated no elastick air, in Exper. 52, 61, 66. Whence 'tis plain, the acid vapours in the air only float in it like the watry vapours; and when strongly attracted by the elastick particles of the air, they firmly adhere to them, and make Salts.

Thus in Experiment 73 we see by the vast quantity of air there is found in *Tartar*, that tho' it contains the other principles of vegetables, yet air with some volatile Salt seems to make up a considerable part of its composition; which air, when by the action of fire it is more firmly united with the earth, and acid sulphureous particles, requires a more intense degree of heat, to extricate it from those adhering substances, as we find in the distillation of *Sal Tartar*, Exper. 74. which Air and volatile Salt are most readily separated by fermentation.

And by Experiment 72, plenty of air arises also from *Nitre*, at the same time that the acid spirit is separated from it by the action of fire.

We find also by Experiment 71, that some air is by the same means obtained from common sea Salt, tho' not in so great plenty, nor so easily, as from *Tartar* and *Nitre*, it being a more fixt body, by reason of the sulphur which abounds in it; neither is it so easily changed in animal bodies, as other Salts are, yet since it fertilizes ground, it must needs be changed by vegetables.

There is good reason also to suspect, that these acid spirits are not wholly free from air particles, notwithstanding there were no elastick ones produced, when they were put into a brisk motion, by the action of fire in Experiment 75. which might be occasioned by the great quantity of acid spirit, in which they were involved. For we see in Experiment 90, that when the acid spirit of *Aqua Regia* was more strongly attracted by the dissolving gold, than by the air particles, then plenty of air particles, which were thus freed from the acid spirit, did continually arise from the *Aqua Regia*, and not from the gold, at least not from the metallick particles of the gold, for that loses nothing of its weight in the solution; so that if any does arise from the gold, it must be what may be latent in the pores of the gold. Whence it is probable, that the air which is obtained by the fermenting mixture of acid and alkaline substances may not arise wholly from the dissolved alkaline body, but in part also from the acid. Thus the great quantity of elastick air, which in Exper. 83. is generated from the mixture of Vinegar and Oystershell, may as well arise in part from the *Tartar*, to which Vinegar owes its acidity, as from the dissolved Oystershell. And what makes it further probable is, that the Vinegar loses its acidity in the ferment, that is its *Tartar*: for dissolving menstruums are generally observed to be changed in fermentation, as well as the dissolved body.

Have we not reason also hence to conclude, that the energy of acid spirits may in some measure be owing to the strongly attracting air particles in them; which active principles may give an impetus to the acid *spiculæ*, as well as the earthy oily matter, which is found in these acid spirits?

There are we see also great store of air particles found in the Analysis of the blood, which arises doubtless as well from the *serum* as from the *crassamentum*, for all the animal fluids and solids have air, and sulphur in them: Which strongly attracting principles seem to be more intimately united together in the more perfect and elaborate part of it, its red globules; so that we may not unreasonably conclude, that air is a band of union here, as well as in Salts: And accordingly we find the greatest plenty of air in the most solid parts of the body, where the cohesion of the parts is the strongest: For by comparing Experiment 49 and 51. we see that much more air was found in the distillation of horn than of blood. And the cohesion of animal substances was not, as we find by the same Experiment, dissolved even in the blood, without considerable violence of fire; tho' it is sometimes done to a fatal degree in our blood, by that more subtile dissolvent fermentation: But we may observe, that volatile Salts, Spirits, and sulphureous Oil, which are at the same time separated from these substances, will not make elastick air.

Experiment CXX.

As elastick air is thus generated by the force of fire, from these and many other substances; so is the elasticity of the air greatly destroyed by sulphureous bodies. Sir *Isaac Newton* observes, "that as light acts upon sulphur, so since all action is mutual, sulphurs ought to act most upon light." And the same may be observed of air and sulphur; for by Experiment 103, it is found that burning sulphur, which is a very strongly attracting substance, powerfully attracts and fixes the elastick particles of air; so that there must needs be a good quantity of un-elastick air particles in oil and flower of sulphur: The first of which is made by burning sulphur under a bell, the other by sublimation: In further confirmation of this it is observed, that *Oleum Sulphuris per Campanam* is with more difficulty made in a dry than a moist air; and I have found by *Experiment* purposely made, that a Candle which burnt 70″ in a very dry receiver, burnt but 64″ in the same receiver, when filled with the fumes of hot water; and yet absorbed one fifth part more air, than when it burnt longer in the dry air.

Sulphur not only absorbs the air when burning in a homogeneal mass, but also in many fermenting mixtures; and as Sir *Isaac Newton* observed the attractive and refractive power of bodies to be greater or less, as they partook more or less of sulphureous oily particles; so there is good reason from these Experiments to attribute the fixing of the elastick particles of the air to the strong attraction of the sulphureous particles with which he says it's probable that all bodies abound more or less.

That great plenty of air is united with sulphur in the oil of vegetables, is evident from the quantity of air that arose from the distillation of oils of Anniseeds and Olives, in Experiment 62. When by fermentation the constituent parts of a vegetable are separated, part of the air flies off in fermentation into an elastick state; part unites with the essential Salt, Water, Oil and Earth, which constitute the Tartar which adhere to the sides of the vessel; the remainder which continues in the fermented liquor, is there, some of it, in a fix'd, and some in an elastick state, which gives briskness to the liquor; their expanding bubbles rising of a very visible size when the weight of the incumbent air is taken off the liquor in a *vacuum*.

And as there was found a greater quantity of air in the deer's horn, than in blood; we may also observe it to be in a much greater proportion in the more solid parts of vegetables, than in their fluid: For we find in Experiment 55. 57. and 60. that near one third part of the substance of the Pease, heart of Oak and Tobacco, were by the action of fire changed from an un-elastick state, to an elastick air: And since a much greater proportion of air is found in the solid than in the fluid parts of bodies; may we not with good reason conclude, that it is very instrumental, as a band of union in those bodies, "Those particles (as Sir *Isaac Newton* observes) receding from one another with the greatest repulsive force, and being most difficultly brought together, which upon contact cohere most strongly? *qu.* 31." And if the attraction of cohesion of an un-elastick air particle be proportionable to its repulsive force in an elastick state; then since its elastick force is found to be so vastly great, so must that of its cohesion be also. Sir *Isaac Newton* calculates from the inflection of the rays of light, that the attracting force of particles, near the point of contact, is 10000, 0000, 0000, 0000 greater than the force of gravity.

Sulphur in a quiescent fix'd state in a large body does not absorb the elastick air, for a hard roll of Brimstone does not absorb air: But when some of that Brimstone, by being powdered and mixt with filings of *Iron*, is set a fermenting, and thereby reduced into very minute particles, whose attraction increases, as their size decreases; then it absorbs elastick air vigorously: As may be seen in many instances under Experiment 95.

The *Walton* mineral, in which there is a good quantity of sulphur, did, when compound *Aqua-fortis* was pour'd on it, in Experiment 96, make a considerable fermentation, and absorb a great quantity of elastick air: But when the ferment was much increased, by adding an equal quantity of water to the like mixture, then instead of absorbing 85 cubick inches as before, it generated 80 cubick inches of air: So that fermenting mixtures, which have sulphur in them, do not always absorb, but sometimes generate air: The reason of which in the Experiment now under consideration seems to be this, *viz.* in the first case a good quantity of elastick air was generated, by the intestine motion of the fermenting ingredients; but there arising thence a thick, acid, sulphureous fume, this fume absorbed a greater quantity of elastick air than was before generated: And we find by Experiment 103 that the sulphureous particles which fly off in the air, do by their attraction destroy its elasticity; for in that Experiment burning Brimstone greatly destroyed the air's elasticity; which must be done by the flame, and ascending fumes; because in the burning of any quantity of Brimstone, the whole mass is in a manner wasted, there remaining only a very little dry Earth: And therefore the absorbed air cannot remain there, but must be absorbed by the ascending fumes which then attract most strongly, when reduced *ad minima*: And 'tis well known that a Candle in burning flies all off into flame and vapour, so that what air it absorbs must be by those fumes.

Experiment CXXI.

And further I have found that these fumes destroy the air's elasticity, for many hours after the Brimstone Match, which made them, was taken out of the vessel, *z z a a*: (Fig. 35.) Those fumes being first cooled by immersing that vessel and its cistern *x x*, or an inverted wine Flask, full of the fumes, under cold water for some time; then marking the surface of the water *z z* I immersed the vessels in warm water: And when all was cold again the following day, I found a good quantity of the air's elasticity was destroyed by the water's ascending above *z z*. And the event was the same upon frequent repetitions of the same Experiment.

But if instead of the fumes of burning Brimstone, I filled a Flask full of fumes from the smoak of wood, after it had done flaming, then there was but half as much air absorbed by those fumes, as there was by the fumes of Brimstone; *viz*, because the smoak of wood was much diluted with the watry vapour which ascended with it out of the wood. And this is doubtless the reason why the smoak of wood, tho' it incommodes the lungs, yet it will not suffocate like that of Charcoal, which is withal more sulphureous, without any mixture of watry vapours.

And that new generated elastick air is resorbed by these fumes, I found by attempting to fire a Match of Brimstone with a burning glass, by means of a pretty large piece of Brown Paper which had been dipped in a strong solution of Nitre, and then dryed: Which Nitre in detonizing generated near two quarts of air, which quantity of air, and a great deal more, was absorbed, when the Brimstone took fire and flam'd vigorously.

So that the 85 cubick inches of air, Experiment 96, which I found upon measuring was absorbed by the *Walton* mineral and compound *Aqua-fortis*, was the excess of what was absorbed by those fumes above what was generated by the fermenting mixture.

And the reason is the same in filings of Iron and Spirit of Nitre, Experiment 94, which also absorbed more than they generated, whether with or without water: The reason of which will appear presently.

Hence also we see the reason why filings of Iron and compound *Aqua-fortis* in the same 94 Experiment absorbed air; and why when mixed with an equal quantity of water it mostly absorbed, but did sometimes generate, and then absorb again: And it was the same with Oil of Vitriol, filings of Iron and Water, and *New-castle* Coal and compound *Aqua-fortis* and others: *viz.* At first, when the ferment was brisk, the absorbing fumes rose fastest, whereby more air was absorbed than generated; but as the ferment abated, to such a degree as to be able still to generate elastick air, but not to send forth a proportionable quantity of fumes, in that case more air would be generated than absorbed.

And in Experiment 95, there are several instances of the air's being in like manner absorbed in lesser degrees, by other fermenting mixtures: As in the mixture of Spirit of Harts-horn with filings of Iron, and with filings of Copper: And Spirit of *Sal Ammoniac* with filings of Copper; and also filings of Iron and Water; powdered Flint and Compound *Aqua-fortis*; powdered *Bristol Diamond* with the same liquor.

It is probable from Experiment 103 and 106, where it was found that the thicker the fuliginous vapours were, the faster they absorbed the air, that if the above-mentioned fermenting mixtures had not been confined in close vessels, but in the open air, where the vapours would have been less dense, that in that case much less air would have been absorbed, perhaps a great deal less than was generated.

In the second case of the *Walton* mineral, Experiment 96, when instead of absorbing, it generated air, the parts of the Compound *Aqua fortis* were then more at liberty to act by being diluted with an equal quantity of water; whereby the ferment being more violent, the particles which constituted the new elastick air were thereby thrown off in greater plenty, and perhaps with a greater degree of elasticity, which might carry them beyond the sphere of attraction of the sulphureous particles.

This is further illustrated by Experiment 94, where filings of Iron and oil of Vitriol alone generated very little; but the like quantities of filings of Iron, with an equal quantity of water, generated 43 cubick inches of air; and the like ingredients, with three times that quantity of water, generated 108 cubick inches.

And tho' the quantity of the ascending fumes (which was in this case of the *Walton* mineral very great) must needs in their ascent absorb a good deal of elastick air, for they will absorb air; yet if where the ferment was so much greater, more elastick air was generated by the fermenting mixture than was absorbed by the ascending fumes; then the quantity of new generated air, which I found between *z z* and *a a*, (Fig. 35.) when I measured it, was equal to the excess of what was generated above what was absorbed.

And probably in this case the air was not absorbed so much in proportion to the density of the fumes as in the first case; because here the sulphureous fumes were much blended with watry vapours: For we find in Experiment 97, that six times more was wasted in fumes in this case than in the others; and therefore probably a good part of the cubick inch of water ascended with the vapour, and might thereby weaken its absorbing power: For watry vapours do not absorb elastick air as the sulphureous ones do; tho' by

Experiment 120, a Candle absorbed more in a damp than in a dry air.

And 'tis from these diluting watry vapours that filings of Iron with Spirit of Nitre and Water, absorbed less than with Spirit of Nitre alone, for in both cases it absorbs more than it generates.

Thus also oil of Vitriol and Chalk generate air, their fume being small, and that much diluted with the watry vapours in the Chalk.

But Lime with oil of Vitriol, or White-Wine Vinegar or Water, make a considerable fume, and absorb good quantities of air: Lime alone left to slaken gradually, as it makes no fume, so it absorbs no air.

We see in Experiment 92, where the ferment was not very sudden nor violent, nor the quantity of absorbing fumes large, that the Antimony and *Aqua-fortis* generated a quantity of air equal to 520 times the bulk of the Antimony; thus also in the mixture of *Aqua-regia* and Antimony, in Experiment 91, while at first the ferment was small, then air was generated; but when with the increasing ferment plenty of fumes arose, then there was a change from a generating to an absorbing state.

Since we find such great quantities of elastick air generated in the solution of animal and vegetable substances; it must needs be that a good deal does constantly arise, from the dissolving of these aliments in the stomach and bowels, which dissolution it greatly promotes: Some of which may very probably be re-sorbed again, by the fumes which arise with them; for we see in Experiment 83 that Oyster-shell and Vinegar, Oyster-shell and Rennet, Oyster-shell and Orange juice, Rennet alone, Rennet and Bread, first generated and then absorbed air; but Oyster-shell with some of the liquor of a Calve's stomach which had fed much upon Hay, did not generate air; and it was the same with Oyster-shell and Ox gall, and spittle, and urine; Oyster-shell and Milk generated a little air, but Limon juice and Milk did at the same time absorb a little: Thus we see that the variety of mixtures in the stomach appear sometimes to generate, and sometimes to absorb air; that is, there is sometimes more generated than absorbed, and sometimes an equal quantity, and sometimes less according to the proportion the generating power of the dissolving aliments bears to the absorbing power of the fumes which arise from them. In a true kindly digestion, the generating power exceeds the absorbing power but a little: But whenever the digestion deviates in some degree from this natural state, to generate a greater proportion of elastick air, then are we troubled more or less with distending *Flatus's*. I had intended to make these and many more Experiments relating to the nature of digestion in a warmth equal to that of the stomach, but have been hitherto prevented by pursuing other Experiments.

Thus we see that all these mixtures do in fermentation generate elastick air, but those which emit thick fumes, charged with sulphur, resorb more than was generated in proportion to the sulphureousness and thickness of those fumes.

I have also shewn in many of the foregoing Experiments, that plenty of true permanent elastick air is generated from the fermenting mixtures of acid and alkaline substances, and especially from the fermentation and dissolution of animal and vegetable bodies: Into whose substances we see it is in a great proportion intimately and firmly incorporated; and consequently, great quantities of elastick air must be continually expended in their production, part of which does we see resume its elastick quality, when briskly thrown off from those bodies by fermentation, in the dissolution of their texture. But part may probably never regain its elasticity, or at least not in many centuries, that especially which is incorporated into the more durable parts of animals and vegetables. However we may with pleasure see what immense treasures of this noble and important element, endued with a most active principle, the all-wise Providence of the great Author of nature has provided; the constant waste of it being abundantly supplyed by heat and fermentation from innumerable dense bodies; and that probably from many of those bodies, which when they had their ascending fumes confined in my Glasses, absorbed more air than they generated, but would be in a more free, open space generate more than they absorbed.

I made some attempts both by fire, and also by fermenting and absorbing mixtures, to try if I could deprive all the particles of any quantity of elastick air of their elasticity, but I could not effect it: There is therefore no direct proof from any of these Experiments, that all the elastick air may be absorbed, tho' tis very probable it may, since we find it in such great plenty generated and absorbed; it may well therefore be all absorbed and changed from an elastick to a fixt state: For as Sir ISAAC NEWTON observes of light, "that nothing more is requisite for producing all the variety of colours, and degrees of refrangibility, than that the rays of light be bodies of different sizes; the least of which may make the weakest and darkest of the colours, and be more easily diverted, by refracting surfaces from the right course; and the rest, as they are bigger and bigger, may make the stronger and more lucid colours—and be more and more difficultly diverted. *Qu.* 29. So *Qu.* 30, he observes of air, that dense bodies by fermentation rarify into several sorts of air, and this air, by fermentation, and sometimes without, returns into dense bodies." And since we find in fact from these Experiments, that air arises from a great variety of dense bodies, both by fire and fermentation, it is probable that they may have very different degrees of elasticity, in proportion to the different size and density of its particles, and the different force with which they were thrown off into an elastick state. "Those particles (as Sir ISAAC NEWTON observes) receding from one another, with the greatest repulsive force, and being most difficultly brought together, which upon contact cohere most strongly." Whence those of the weakest elasticity, will be least able to resist a counter-acting power, and will therefore be soonest changed from an elastick to a fixt state. And 'tis consonant to reason to think, that the air may consist of infinite degrees of these, from the most elastick and repelling, till we come to the more sluggish, watry and other particles, which float in the air; yet the repelling force of the least elastick particle, near the surface of the earth, while it continues in that elastick state, must be superior to the incumbent pressure of a column of air, whose height is equal to that of the atmosphere, and its base to the surface of the sphere of its elastick activity.

Thus upon the whole, we see that air abounds in animal, vegetable and mineral substances; in all which it bears a considerable part: if all the parts of matter were only endued with a strongly attracting power, whole nature would then immediately become one unactive cohering lump; wherefore it was absolutely necessary, in order to the actuating and enlivening this vast mass of attracting matter, that there should be every where intermixed with it a due proportion of strongly repelling elastick particles, which might enliven the whole mass, by the incessant action between them and the attracting particles: And since these elastick particles are continually in great abundance reduced by the power of the strong attracters, from an elastick, to a fixt state; it was therefore necessary that these particles should be endued with a property of resuming their elastick state, whenever they were disengaged from that mass, in which they were fixed; that thereby this beautiful frame of things might be maintained, in a continual round of the production and dissolution of animal and vegetable bodies.

The air is very instrumental in the production and growth of animals and vegetables, both by invigorating their several juices while in an elastick state, and also by greatly contributing in a fix'd state to the union and firm connection of the several constituent parts of those bodies, *viz.* their water, salt, sulphur and earth. This band of union, in conjunction with the external air, is also a very powerful agent in the dissolution and corruption of the same bodies, for it makes one in every fermenting mixture; the action and re-action of the aereal and sulphureous particles is in many fermenting mixtures so great, as to excite a burning heat, and in others a sudden flame: And it is we see by the like action and re-action of the same principles, in fuel and the ambient air, that common culinary fires are produced and maintained.

Tho' the force of its elasticity is so great as to be able to bear a prodigious pressure, without losing that elasticity, yet we have from the foregoing Experiments evident proof, that its elasticity is easily, and in great abundance destroyed; and is thereby reduced to a fixt state, by the strong attraction of the acid sulphureous particles, which arise either from fire or from fermentation: And therefore elasticity is not an essential immutable property of air particles; but they are, we see, easily changed from an elastick to a fixt state, by the strong attraction of the acid, sulphureous and saline particles which abound in the air. Whence it is reasonable to conclude, that our atmosphere is a *Chaos*, consisting not only of elastick, but also of unelastick air particles, which in great plenty float in it, as well as the sulphureous, saline, watry and earthy particles, which are no ways capable of being thrown off into a permanently elastick state, like those particles which constitute true permanent air.

Since then air is found so manifestly to abound in almost all natural bodies; since we find it so operative and active a principle in every chymical operation, since its constituent parts are of so durable a nature, that the most violent action of fire, or fermentation, cannot induce such an alteration of its texture, as thereby to disqualify it from resuming, either by the means of fire, or fermentation, its former elastick state; unless in the case of vitrification, when with the vegetable Salt and Nitre, in which it is incorporated, it may perhaps some of it with other chymical principles be immutably fixt: Since then this is the case, may we not with good reason adopt this now fixt, now volatile *Proteus* among the chymical principles, and that a very active one, as well as acid sulphur; notwithstanding it has hitherto been overlooked and rejected by Chymists, as no way intitled to that denomination?

If those who unhappily spent their time and substance in search after an imaginary production, that was to reduce all things to gold, had, instead of that fruitless pursuit, bestowed their labour in searching after this much neglected volatile *Hermes*, who has so often escaped thro' their burst receivers, in the disguise of a subtile spirit, a mere flatulent explosive matter; they would then instead of reaping vanity, have found their researches rewarded with very considerable and useful discoveries.

CHAP. VII.

Of Vegetation.

We are but too sensible, that our reasonings about the wonderful and intricate operations of nature are so full of uncertainty, that as the wise-man truly observes, *hardly do we guess aright at the things that are upon earth, and with labour do we find the things that are before us.* Wisdom Chap. ix. *v.* 16. And this observation we find sufficiently verified in vegetable nature, whose abundant productions, tho' they are most visible and obvious to us, yet are we much in the dark about the nature of them, because the texture of the vessels of plants is so intricate and fine, that we can trace but few of them, tho' assisted with the best microscopes. We have however good reason to be diligent in making farther and farther researches; for tho' we can never hope to come to the bottom and first principles of things, yet in so inexhaustible a subject, where every the smallest part of this wonderful fabrick is wrought in the most curious and beautiful manner, we need not doubt of having our inquiries rewarded, with some further pleasing discovery; but if this should not be the reward of our diligence, we are however sure of entertaining our minds after the most agreeable manner, by seeing in every thing, with surprising delight, such plain signatures of the wonderful hand of the divine architect, as must necessarily dispose and carry our thoughts to an act of adoration, the best and noblest employment and entertainment of the mind.

What I shall here say, will be chiefly founded on the following experiments; and on several of the preceding ones, without repeating what has already been occasionally observed on the subject of vegetation.

We find by the chymical analysis of vegetables, that their substance is composed of sulphur, volatile salt, water and earth; which principles are all endued with mutually attracting powers, and also of a large portion of air, which has a wonderful property of strongly attracting in a fixt state, or of repelling in an elastick state, with a power which is superior to vast compressing forces, and it is by the infinite combinations, action and re-action of these principles, that all the operations in animal and vegetable bodies are effected.

These active aereal particles are very serviceable in carrying on the work of vegetation to its perfection and maturity. Not only in helping by their elasticity to distend each ductile part, but also by enlivening and invigorating their sap, where mixing with the other mutually attracting principles they are by gentle heat and motion set at liberty to assimilate into the nourishment of the respective parts: "The soft and moist nourishment easily changing its texture by gentle heat and motion, which congregates homogeneal bodies, and separates heterogeneal ones." *Newton's Opticks*, qu. 31. The sum of the attracting power of these mutually acting and re-acting principles being, while in this nutritive state, superior to the sum of their repelling power, whereby the work of nutrition is gradually advanced by the nearer and nearer union of these principles, from a lesser to a greater degree of consistency, till they are advanced to that viscid ductile state, whence the several parts of vegetables are formed; and are at length firmly compacted into hard substances, by the flying off of the watry diluting vehicle; sooner or later, according to the different degrees of cohesion of these thus compacted principles.

But when the watry particles do again soak into and disunite them, and their repelling power is thereby become superior to their attracting power; then is the union of the parts of vegetables thereby so thoroughly dissolved, that this state of putrefaction does by a wise order of Providence fit them to resuscitate again, in new vegetable productions; whereby the nutritive fund of nature can never be exhausted: Which being the same both in animals and vegetables, it is thereby admirably fitted by a little alteration of its texture to nourish either.

Now, tho' all the principles of vegetables are in their due proportion necessary to the production and perfection of them; yet we generally find greater proportions of Oil in the more elaborate and exalted parts of vegetables: And thus Seeds are found to abound with Oil, and consequently with sulphur and air, as we see by Exper. 56, 57, 58. which Seeds containing the rudiments of future vegetables, it was necessary that they should be well stored with principles that would both preserve the Seed from putrefaction, and also be very active in promoting germination and vegetation. Thus also by the grateful odours of flowers we are assured, that they are stored with a very subtile, highly sublimed Oil, which perfumes the ambient air, and the same may be observed from the high tastes of fruits.

And as Oil is an excellent preservative against the injuries of cold, so it is found to abound in the sap of the more northern trees; and it is this which in ever-greens keeps their leaves from falling.

But plants of a less durable texture, as they abound with a greater proportion of Salt and Water, which is not so strongly attracting as sulphur and air, so are they less able to endure the cold; and as plants are observed to have a greater proportion of Salt and Water in them in the spring, than in the autumn, so are they more easily injured by cold in the spring, than in a more advanced age, when their quantity of oil is increased, with their greater maturity.

Whence we find that nature's chief business, in bringing the parts of a vegetable, especially its fruit and seed to maturity, is to combine together in a due proportion, the more active and noble principles of sulphur and air, that chiefly constitute oil, which in its most refined state is never found without some degree of earth and salt in it.

And the more perfect this maturity is, the more firmly are these noble principles united. Thus Rhenish Wines, which grow in a more northern climate, are found to yield their Tartar, *i. e.* by Exper. 73. their incorporated air and sulphur in greater plenty, than the stronger Wines of hotter countries, in which these generous principles are more firmly united: And particularly in *Madera* Wine, they are fixt to such a degree, that that Wine requires a considerable degree of warmth, such as would soure many other Wines, to keep it in order, and give it a generous taste; and 'tis from the same reason, that small *French* Wines are found to yield more spirit in distillation, than strong *Spanish Wines*.

But when, on the other hand, the crude watry part of the nutriment bears too great a proportion to the more noble principles, either in a too luxuriant state of a plant, or when its roots are planted too deep, or it stands in too shady a position, or in a very cold and wet summer; then it is found, that either no fruit is produced, or if there be any, yet it continues in a crude watry state; and never comes to that degree of maturity, which a due proportion of the more noble principles would bring it to.

Thus we find in this, and every other part of this beautiful scene of things, when we attentively consider them, that the great Author of nature has admirably tempered the constituent principles of natural bodies, in such due proportions as might best fit them for the state and purposes they were intended for.

It is very plain from many of the foregoing Experiments and Observations, that the leaves are very serviceable in this work of vegetation, by being instrumental in bringing nourishment from the lower parts, within the reach of the attraction of the growing fruit; which like young animals is furnished with proper instruments to suck it thence. But the leaves seem also designed for many other noble and

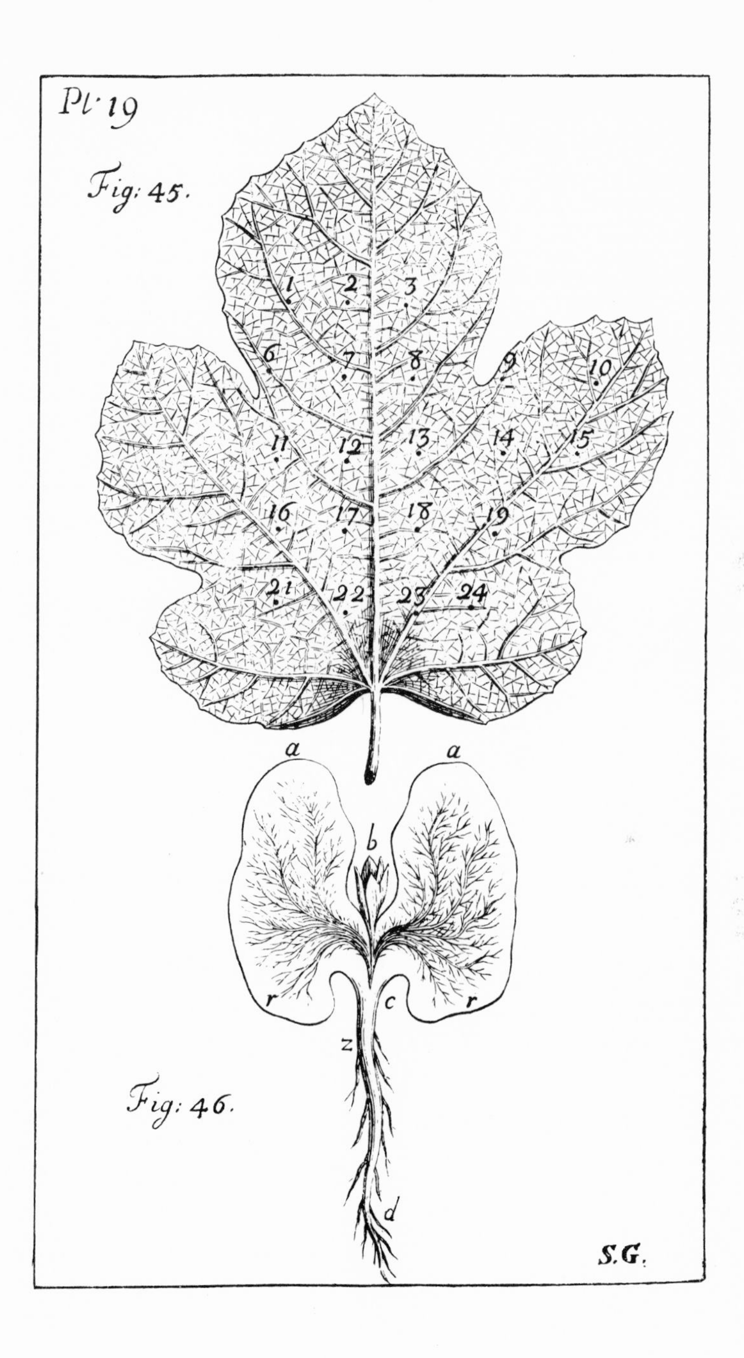
Pl. 19
Fig: 45.
1
2
3
6
7
8
9
10
11
12
13
14
15
16
17
18
19
21
22
23
24
a
a
b
c
r
r
z
Fig: 46.
d
S.G.

important services; for nature admirably adapts her instruments so as to be at the same time serviceable to many good purposes. Thus the leaves, in which are the main excretory ducts in vegetables, separate and carry off the redundant watry fluid, which by being long detained, would turn rancid and prejudicious to the plant, leaving the more nutritive parts to coalesce; part of which nourishment, we have good reason to think, is conveyed into vegetables thro' the leaves, which do plentifully imbibe the Dew and Rain, which contain Salt, Sulphur, *&c.* For the air is full of acid and sulphureous particles, which when they abound much, do by the action and re-action between them and the elastick air cause that sultry heat, which usually ends in lightning and thunder: And these new combinations of air, sulphur and acid spirit, which are constantly forming in the air, are doubtless very serviceable, in promoting the work of vegetation; when being imbibed by the leaves, they may not improbably be the materials out of which the more subtile and refined principles of vegetables are formed: For so fine a fluid as the air seems to be a more proper medium, wherein to prepare and combine the more exalted principles of vegetables, than the grosser watry fluid of the sap; and for the same reason, 'tis likely, that the most refined and active principles of animals are also prepared in the air, and thence conveyed thro' the lungs into the blood; and that there is plenty of these sulphureo-aereal particles in the leaves, is evident from the sulphureous exudations, which are found at the edges of leaves, which Bees are observed to make their waxen cells of, as well as of the dust of flowers: And that wax abounds with sulphur is plain from its burning freely, *&c.*

We may therefore reasonably conclude, that one great use of leaves is what has been long suspected by many, *viz.* to perform in some measure the same office for the support of the vegetable life, that the lungs of animals do, for the support of the animal life; Plants very probably drawing thro' their leaves some part of their nourishment from the air.

But as plants have not a dilating and contracting *Thorax*, their inspirations and expirations will not be so frequent as those of Animals, but depend wholly on the alternate changes from hot to cold, for inspiration, and *vice versâ* for expiration; and tis' not improbable, that plants of more rich and racy juices may imbibe and assimilate more of this aereal food into their constitutions, than others, which have more watry vapid juices. We may look upon the Vine as a good instance of this, which in Exper. 3. perspired less than the Apple-tree. For as it delights not in drawing much watry nourishment from the earth by its roots, so it must therefore necessarily be brought to a more strongly imbibing state at night, than other trees, which abound more with watry nourishment; and it will therefore consequently imbibe more from the air. And likely this may be the reason, why plants in hot countries abound more with fine aromatick principles, than the more northern plants, for they do undoubtedly imbibe more dew.

And if this conjecture be right, then it gives us a farther reason, why trees which abound with moisture, either from too shaded a position, or a too luxurious state are unfruitful, *viz.* because, being in these cases more replete with moisture, they cannot not imbibe so strongly from the air, as others do, that great blessing the dew of Heaven.

And as the most racy generous tastes of fruits, and the grateful odours of flowers, do not improbably arise from these refined aereal principles, so may the beautiful colours of flowers be owing in a good measure to the same original; for it is a known observation, that a dry soil contributes much more to their variegation than a strong moist one does.

And may not light also, by freely entring the expanded surfaces of leaves and flowers, contribute much to the ennobling the principles of vegetables? for Sir *Isaac Newton* puts it as a very probable query, "Are not gross bodies and light convertible into one another? and may not bodies receive much of their activity from the particles of light, which enter their composition? The change of bodies into light, and of light into bodies, is very conformable to the course of nature, which seems delighted with transmutations. *Opt. qu.* 30."

EXPERIMENT CXXII.

That the leaves and plants do imbibe elastick air, I have some reason to suspect from the following Experiment, *viz.* In *May* I set some well rooted plants of spear-mint in two glass cisterns full of water, which cisterns were set on pedestals, and had inverted chymical receivers put over them, as in (Fig. 35.) the water being drawn up to *a a*, half way their necks: In this inclosed moist state the plants looked pretty florid for a month, and made, as I think, some few weak lateral shoots, tho' they did not grow in height; they were not quite dead till after six weeks, when it was found that the water was risen in both glasses from *a a* towards *z z*, in bulk about 20 cubick inches: But as there was not so exact an account taken of the different temperature of the air, as to heat and cold, as there ought to have been, I am not certain, whether that rising of the water might not be owing to a greater coolness of the air at the six weeks end, than when they were first placed under the glasses; and therefore do not depend on this Experiment; but thought it proper to mention it, as well deserving to be repeated with greater accuracy, both with Mint, and other proper plants, by noting the temperature of the air on a *Thermometer*, hanging near the receivers, and observing after some time, whether the water *a a* be risen, notwithstanding the air be no cooler than when the Mint was first placed under the glass. And for greater certainty, it would be adviseable to suspend in the same manner another like receiver with no Mint, but only water in it, up to *a a*.

EXPERIMENT CXXIII.

In order to find out the manner of the growth of young shoots, I first prepared the following instrument, *viz.* I took a small stick *a*, (Fig. 40.) and at a quarter of an inch distance from each other, I run the points of five pins, 1, 2, 3, 4, 5, thro' the stick, so far as to stand $\frac{1}{4}$ of an inch from the stick, then bending down the great ends of the pins, I bound them all fast with waxed thread; I provided also some red lead mixed with oil.

In the spring, when the Vines had made short shoots, I dipped the points of the pins in the paint, and then pricked the young shoot of a Vine, (Fig. 41.) with the five points at once, from *t* to *p*: I then took off the marking instrument, and placing the lowest point of it in the hole *p*, the uppermost mark, I again pricked fresh holes from *p* to *l*, and then marked the two other points *i h*; thus the whole shoot was marked every $\frac{1}{4}$ inch, the red paint making every point remain visible.

(Fig. 42.) shews the true proportion of the same shoot, when it was full grown, the *September* following; where every corresponding point is noted with the same letter.

The distance from *t* to *s* was not enlarged above $\frac{1}{60}$ part of an inch; from *s* to *q*, the $\frac{1}{26}$ of an inch; from *q* to *p*, $\frac{3}{8}$; from *p* to *o*, $\frac{3}{8}$; from *o* to *n*, $\frac{6}{10}$; from *n* to *m* $\frac{9}{10}$ from *m* to *l*, 1 + $\frac{1}{10}$ of an inch; from *l* to *i*, 1 + $\frac{3}{10}$ inch nearly; and from *i* to *h* three inches.

In this Experiment we see that the first joint to *r* extended very little; it being almost hardened, and come near to its full growth, when I marked it: The next joint, from *r* to *n*, being younger, extended something more; and the third joynt from *n* to *k* extended from $\frac{3}{4}$ of an inch to 3 + $\frac{1}{2}$ inches; but from *k* to *h*, the very tender joint, which was but $\frac{1}{4}$ inch long, when I marked it, was when full grown three inches long.

We may observe, that nature in order to supply these young growing shoots with plenty of ductile matter is very careful to furnish at small distances the young shoots of all sorts of trees, with many leaves throughout their whole length, which serve as so many joyntly acting powers placed at different stations, thereby to draw with more ease plenty of sap to the extending shoot.

The like provision has nature made in the Corn, Grass, Cane, and Reed kind; the leafy spires, which draw the nourishment to each joynt, being provided long before the stem shoots, which tender stem in its tender ductile state would most easily break and dry up too soon, so as to prevent its due growth, had not nature to prevent both these inconveniences provided strong *Thecas* or *Scabbards*, which both support and keep long in a supple ductile state the tender extending stem.

I marked in the same manner as the Vine, at the proper seasons, young *Honeysuckle* shoots, young *Asparagus*, and young *Sun-flowers*; and I found in them all a gradual scale of unequal extentions, those parts extending most which were tenderest. The white part of the *Asparagus*, which was under ground, extended very little in length, and accordingly we find the fibres of the white part very tough and stringy: But the greatest extension of the tender green part, which was about 4 inches above the ground when I marked it, separated the marks from a quarter of an inch, to twelve inches distance; the greatest distention of the *Sunflower* was from $\frac{1}{4}$ inch, to four inches distance.

From these Experiments, it is evident, that the growth of a young bud to a shoot consists in the gradual dilatation and distention of every part, they are extended to their full each other in the bud, as may plainly and distinctly be seen in the slit bud of the Vine and Fig-tree; but by this gradual distention of every part, they are extended to their full length. And we may easily conceive how the longitudinal capillary tubes still retain their hollowness, notwithstanding their being distended, from the like effect in melted glass tubes, which retain a hollowness, tho' drawn out to the finest thread.

The whole progress of the first joynt *r* is very short in comparison of the other joynts; because, at first setting out its leaves being very small, and the season then cooler than afterwards; 'tis probable, that but little sap is conveyed to it; and therefore it extending but slowly, its fibres are in the mean time grown tough and hard, before it can arrive to any considerable length. But as the season advances, and the leaves inlarge, greater plenty of nourishment being thereby conveyed, the second joynt grows longer than the first, and the 3d and 4th still on gradually longer than the preceding; these do therefore in equal times make greater advances than the former.

The wetter the season, the longer and larger shoots do vegetables usually make; because their soft ductile parts do then continue longer in a moist, tender state; but in a dry season the fibres sooner harden, and stop the further growth of the shoot; and this may probably be one reason why the two or three last joynts of every shoot are usually shorter than the middle joynts; *viz*, because they shooting out in the more advanced hot dry summer season, their fibres are soon hardened and dryed, and are withal checked in their growth by the cool autumnal nights: I had a vine shoot of one year's growth which was 14 feet long, and had 39 joynts, all pretty nearly of an equal length, except some of the first and last.

And for the same reason, Beans and many other plants, which stand where they are much shaded, being thereby kept continually moist, do grow to unusual heights, and are drawn up as they call it by the over shadowing Trees, their parts being kept long, soft and ductile: But this very moist shaded state is usually attended with sterility; very long joynts of vines are also observed to be unfruitful.

This Experiment, which shews the manner of the growth of shoots, confirms *Borelli's* opinion, who in his Book *De motu Animalium*, part second Chap. 13, supposes the tender growing shoots to be distended like soft wax by the expansion of the moisture in the spongy pith; which dilating moisture, he with good reason concludes is hindered from returning back, while it expands, by the sponginess of the pith, without the help of valves. For 'tis very probable that the particles of water, which immediately adhere to, and are strongly imbibed into, and attracted by every fibre of the spongy pith, will suffer some degree of expansion before they can be detached by the sun's warmth from each attracting fibre, and consequently the mass of spongy fibres, of which the pith consists, must thereby be extended.

And that the pith may be the more serviceable for this purpose, nature has provided in most shoots a strong partition at every knot, which partitions serve not only as plinths, or abutments for the dilating pith to exert its force on, but also to prevent the rarified sap's too free retreat from the pith.

But a dilating spongy substance, by equally expanding it self every way, would not produce an oblong shoot, but rather a globose one, like an Apple; to prevent which inconvenience we may observe, that nature has provided several Diaphragms, besides those at each knot, which are placed at small distances across the pith; thereby preventing its too great lateral dilatation. These are very plain to be seen in Walnut-tree shoots; and the same we may observe in the pith of the branches of the sun-flower, and of several other plants; where tho' these Diaphragms are not to be distinguished while the pith is full and replete with moisture, yet when it drys up, they are often plain to be seen; and it is further observed, that where the pith consists of distinct vesicles, the fibres of those vesicles are often found to run horizontally, whereby they can the better resist the too great lateral dilatation of the shoot.

We may observe that nature makes use of the same artifice, in the growth of the feathers of Birds, which is very visible in the great pinion feathers of the wing, the smaller and upper part of which is extended by a spongy pith, but the lower and bigger quill part, by a series of large vesicles, which when replete with dilating moisture do extend the quill, but when the quill is full grown, the vesicles are always dry; in which state we may plainly observe every vesicle to be contracted at each end by a Diaphragm or Sphincter, whereby its too great lateral dilatation is prevented, but not its distention lengthwise.

And as this pith in the quill grows dry and useless after the quill is full grown, we may observe the same in the pith of trees, which is always succulent and full of moisture while the shoot is growing, by the expansion of which the tender ductile shoot is distended in every part, its fibres being at the same time kept supple by this moisture; but when each year's shoot is full grown, then the pith gradually drys up, and continues for the future dry and kiksey, its vesicles being ever after empty; nature always carefully providing for the succeeding year's growth by preserving a tender ductile part in the bud replete with succulent pith.

And as in vegetables, so doubtless in animals, the tender ductile bones of young animals are gradually increased in every part, that is not hardened and ossified; but since it was inconsistent with the motion of the joynts to have the ends of the bones soft and ductile as in vegetables; therefore nature makes a wonderful provision for this at the glutinous serrated joyning of the heads to the shanks of the bones; which joyning while it continues ductile the animal grows, but when it ossifies then the animal can no longer grow. As I was assured by the following Experiment, *viz*. I took a half-grown Chick, whose leg-bone was then two inches long, and with a sharp pointed Iron at half an inch distance I pierced two small holes through the middle of the scaly covering of the leg, and shin-bone; two months after I killed the Chick, and upon laying the bones bare, I found on it obscure remains of the two marks I had made at the same distance of half an inch: So that that part of the bone had not at all distended lengthwise, since the time that I marked it: Notwithstanding the bone was in that time grown an inch more in length, which growth was mostly at the upper end of the bone, where a wonderful provision is made for its growth at the joining of its head to the shank, called by Anatomists Symphysis.

And as the bones grow in length and size; so must the membranous, the muscular, the nervous, the cartilaginous and vascular fibres of the animal body necessarily extend and expand, from the ductile nutriment which nature furnishes every part withal; in which respects animal bodies do as truly vegetate as do the growing vegetables. Whence it must needs be of the greatest consequence, that the growing animal be supplied with proper nourishment for that purpose, in order to form a strong athletick constitution: For when growing nature is deprived of proper materials for this purpose, then is she under a necessity of drawing out very slender threads of life, as is too often the case of young growing persons, who by indulging in spirituous liquors, or other excesses, do thereby greatly deprave the nutritive ductile matter, whence all the distending fibres of the body are supplied.

Since we are by these Experiments assured that the longitudinal fibres, the sap vessels of wood in its first year's growth, do thus distend in length by the extension of every part; and since nature in similar productions makes use of the same or nearly the same methods: These considerations make it not unreasonable to think, that the second and following years additional ringlets of wood are not formed by a meerly horizontal dilatation of the vessels; for it is not easy to conceive, how longitudinal fibres and tubular sap-vessels should thus be formed; but rather by the shooting of the longitudinal fibres lengthways under the bark as young fibrous shoots of roots do, in the solid Earth. The observations on the manner of the growth of the ringlets of wood in Experiment 46 (Fig. 30.) do further confirm this.

I intended to have made father researches into this matter by proper Experiments, but have not yet found time for it.

But whether it be by an horizontal or longitudinal shooting, we may observe that nature has taken great care to keep the parts between the bark and wood always very supple with slimy moisture, from which ductile matter the woody fibres, vesicles and buds are formed.

Thus we see that nature, in order to the production and growth of all the parts of animals and vegetables, prepares her ductile matter: In doing of which she selects and combines particles of very different degrees of mutual attraction, curiously proportioning the mixture according to the many different purposes she designs it for; either for bony or more lax fibres of very different degrees in animals, or whether it be for the forming of woody or more soft fibres of various kinds in vegetables.

The great variety of which different substances in the same vegetable prove, that there are appropriate vessels for conveying very different sorts of nutriment. And in many vegetables some of those appropriate vessels are plainly to be seen replete either with milky, yellow, or red nutriment.

Dr. *Keill*, in his account of animal secretion, page 49, observes, that where nature intends to separate a viscid matter from the blood, she contrives very much to retard its motion, whereby the intestine motion of the blood being allayed, its particles can the better coalesce in order to form the viscid secretion. And Dr. *Grew*, before him, observed an instance of the same contrivance in vegetables where a secretion is intended, that is to compose a hard substance, *viz.* in the kernell or feed of hard stone fruits, which does not immediately adhere to, and grow from the upper part of the stone, which would be the shortest and nearest way to convey nourishment to it; but the single umbilical vessel, by which the kernel is nourished, fetches a compass round the concave of the stone, and then enters the kernel near its cone, by which artifice this vessel being much prolonged, the motion of the sap is thereby retarded, and a viscid nutriment conveyed to the seed, which turns to hard substance.

The like artifice of nature we may observe in the long capillary fibrous vessels which lie between the green hull, and the hard shell of the Walnut, which are analogous to the fibrous Mace of Nutmegs, the ends of whose hairy fibres are inserted into the angles of the furrows in the Walnut-shell: Their use is therefore doubtless to carry in those long distinct vessels the very viscous matter which turns, when dry, to a hard shell; whereas were the shell immediately nourished from the soft pulpous hull that surrounds it, it would certainly be of the same soft constitution: The use of the hull being only to keep the shell in a soft ductile state till the Nut has done growing.

We may observe the like effect of a flower motion of the sap in Ever-greens, which perspiring little, their sap moves much more slowly than in more perspiring Trees; and is therefore much more viscid, whereby they are better enabled to outlive the winter's cold. It is observed that the sap of Ever-greens in hot Countries is not so viscous as the sap of more Northern Ever-greens, as the fir, *&c.* for the sap in hotter Countries must have a brisker motion, by means of its greater perspiration.

Experiment CXXIV.

In order to enquire into the manner of the expansion of leaves, I provided a little Oaken board or spatula, *a b c d* of this shape and size, (Fig. 43.) thro' the broad part at a quarter of an inch distance from each other; I run the points of 25 pins *x x* which stood $\frac{1}{4}$ inch thro', and divided a square inch into 16 equal parts.

With this instrument in the proper season, when leaves were very young, I pricked several of them thro' at once, with the points of all these pins, dipping them first in the red lead, which made lasting marks.

(Fig. 44.) represents the shape and size of a young Fig-leaf, when first marked with red points, $\frac{1}{4}$ inch distance from each other.

(Fig. 45.) represents the same full grown leaf, and the numbers answer to the corresponding numbers in the young leaf: Whereby may be seen how the several points of the growing leaf were separated from each other, and in what proportion, *viz.* from a quarter of an inch, to about three quarter's of an inch distance.

In this Experiment we may observe that the growth and expansion of the leaves is owing to the dilatation of the vesicles in every part, as the growth of a young shoot was shewn to be owing to the same cause in the foregoing Experiment; and doubtless the case is the same in all fruits.

If these Experiments on leaves were further pursued, there might probably be many curious observations made in relation to the shape of leaves: By observing the difference of the progressive and lateral motions of these points in different leaves, that were of very different lengths in proportion to their breadths.

That the force of dilating sap and air, included in the innumerable little vesicles of young tender shoots and leaves, is abundantly sufficient for the extending of shoots, and expanding of leaves; we have evident proof from the great force we find in the sap of the Vine, chap. 3d. and from the vast force, with which insinuating moisture expanded the Pease. Experiment 32. we see the great power of expanding water, when heated in the engine to raise water by fire: And water with air and other active particles in capillary tubes, and innumerable small vesicles, do doubtless act with a great force, tho' expanded with no more heat than what the Sun's warmth gives them.

And thus we see that nature exerts a considerable, tho' secret and silent power, in carrying on all her productions; which demonstrates the wisdom of the Author of nature in giving such due proportion and direction to these powers, that they uniformly concur to the production and perfection of natural Beings; whereas were such powers under no guidance, they must necessarily produce a *Chaos*, instead of that regular and beautiful system of nature which we see.

We may plainly see the influence of the Sun's warmth in expanding the sap in all the parts of vegetables, as well in the roots as the body that is above ground, by the influence it has on the six *Thermometers* described under *Experiment* 20, five of which were fixed at different depths from two inches, to two feet under ground, the other being exposed to the open air.

When the greatest noon tide heat the spirit of that which was exposed to the Sun was risen, since the early morning, from 21 to 48 degrees; then the spirit in the second *Thermometer*, whose ball was two inches under ground, was at 45 degrees, and the 3d, 4th, and 5th *Thermometers* were gradually of less and less degrees of heat, as they were placed lower in the ground to the sixth *Thermometer*, which was two feet under ground, in which the spirit was 31 degrees high. In this state of heat on all the parts of the vegetable, we see the Sun must have a very considerable influence in expanding the sap in all its parts. The warmth was much greater on the body above ground, than on the roots which were two feet deep; those roots, and parts of roots which are deepest, as they feel much less of the Sun's warmth, so are they not so soon, nor so much affected by the alternacies of day and night, warm and cold: but that part of vegetables, which is above ground, must have its sap considerably rarified, when the heat increased from morning to two a clock afternoon, so much as to raise the spirit in the 1st *Thermometer* from 21 to 48 degrees above the freezing point.

When in the coldest days of the winter 1724, the frost was so intense as to freeze the surface of stagnant water near an inch thick, then the spirit in the *Thermometer* which was exposed to the open air, was fallen four degrees below the freezing point; the spirit of that whose ball was two inches under ground, was four degrees above the freezing point; the 3d, 4th and 5th *Thermometers* were proportionably fallen less and less, as they were deeper, to the 6th *Thermometer*, which being two feet under ground, the spirit was 10 degrees above the freezing point. In this state of things the work of vegetation seemed to be wholly at a stand, at least within the reach of the frost.

But when the cold was so far relaxed, as to have the spirit in the first *Thermometer* but 5 degrees above the freezing point, the second 8 degrees, and the 6th 13 degrees, tho' it was still very cold, yet this being some advance from freezing towards warm, and there being consequently some expansion of the sap, several of the hardy vegetables grew, *viz.* some Ever-greens, Snow-drops, Crocus's, *&c.* which forward hardy plants do probably partake much of the nature of Ever-greens in perspiring little; and the motion of their sap being consequently very slow, it will become more viscous, as in Ever-greens; and thereby the better able to resist the winter's cold: And the small expansive force, which this sap acquires in the winter, is mostly exerted in extending the plant, little of it being wasted in proportion to the summer's perspiration.

Supported by the evidence of many of the foregoing Experiments, I will now trace the vegetation of a tree from its first seminal plant in the Seed to its full maturity and production of other Seeds, without entring into a particular description of the structure of the parts of vegetables, which has already been accurately done by Dr. *Grew* and *Malpighi*.

We see by Experiment 56, 57, 58, on distilled Wheat, Pease and Mustard-seed, what a wonderful provision nature has made, that the Seeds of Plants should be well stored with very active principles, which principles are there compacted together by him, who curiously adapts all things to the purposes for which they are intended, with such a just degree of cohesion as retains them in that state till the proper season of germination; for if they were of a more lax constitution, they would too soon dissolve like the other tender annual parts of plants: And if they were more firmly connected, as in the heart of Oak, they must necessarily have been many years in germinating, tho' suppled with moisture and warmth.

When a Seed is sown in the ground, in a few days it imbibes so much moisture, as to swell with very great force; as we see in the Experiment on Pease in an iron pot, this forcible swelling of the lobes of the Seed *a r, a r* (Fig. 46.) does probably protrude moisture and nourishment from the capillary vessels *r r*, which are called the Seed roots, into the radicle *c, z, d*, which radicle, when it has shot some length into the ground, does then imbibe nourishment from thence; and after it has acquired sufficient strength, as this tender ductile root is extending from *z* to *c*, it must necessarily carry the expanding Seed-lobes upwards, at the same time that the dilating from *z* to *d* makes it shoot downwards; and when the root is thus far grown, it supplies the Plume *b* with nourishment, which thereby swelling and extending opens the lobes *a r, a r*, which are at the same time raised above ground with the Plume; where they by expanding and growing thinner turn to green leaves, (except the Seeds of the pulse kind) which leaves are of such importance to the yet tender Plume, that it perishes, or will not thrive if they are pulled off; which makes it probable, that they do the same office to the Plume that the leaves adjoyning to Apples, Quinces and other fruits do to them, *viz.* they draw sap within the reach of their attraction; see Exper. 8 and 30. But when the Plume is so far advanced in growth, as to have branches and expanded leaves to draw up nourishment; then these supplemental seminal leaves, *a r, a r*, being of no farther use, do perish; not only because the now grown and more expanded leaves of the young plant or tree, do so overshadow the supplemental leaves, that their former more plentiful perspiration is much abated; and thereby also their power of attracting sap fails; but also because the sap is drawn from them by the leaves, and they being thus deprived of nourishment, do perish.

As the tree advances in stature, the first, second, third, and fourth order of lateral branches shoot out, each lower order being longer than those immediately above them; not only on account of primogeniture, but also because being inserted in larger parts of the trunk, and nearer the root, they have the advantage of being served with greater plenty of sap, whence arises the beautiful parabolical figure of trees.

But when trees stand thick together in Woods or Groves, this their natural shape is altered, because the lower lateral branches being much shaded, they can perspire little; and therefore drawing little nourishment, they perish; but the top branches, being exposed to a free drying air, they perspire plentifully, and thereby drawing the sap to the top, they advance much in height: But *vice versâ*, if when such a Grove of tall trees is cut down, there be left here and there a single tree, that tree will then shoot out lateral branches; the leaves of which branches now perspiring freely, will attract plenty of sap, on which account the top being deprived of its nourishment, it usually dies.

And as trees in a Grove or Wood grow only in length, because all the nourishment is by the leaves drawn to the top, most of the small lateral shaded branches in the mean time perishing for want of perspiration and nutrition; so the case is the very same in the branches of a tree, which usually making an angle of about 45 degrees with the stem of the tree, do thereby beautifully fill up at equal and proper distances the space between the lower branches, and the top of the tree, forming thereby as it were a parabolical Grove or Thicket; which shading the arms, the small lateral shoots of those arms usually perish for want of due perspiration; and therefore the arms continue naked like the bodies of

Trees in a grove; all the nourishment being drawn up to the tops of the several branches by the leaves which are there exposed to the warm sun and free drying air, whereby the branches of Trees expand much.

And where the lateral branches are very vigorous, so as to make strong shoots, and attract the nourishment plentifully, there the tree usually abates in its height: But where the tree prevails in height, as in groves, there commonly its lateral branches are smallest. So that we may look upon a tree as a complicated Engine which has as many different powers as it has arms and branches, each drawing from their common fountain of life the root: And the whole of each yearly growth of the tree will be proportionable to the sum of their attracting powers, and the quantity of nourishment the root affords: But this attracting power and nourishment will be more or less, according to the different ages of the tree, and the more or less kindly seasons of the year.

And the proportional growth of their lateral and top branches, in relation to each other, will much depend on the difference of their several attracting powers. If the perspiration and attraction of the lateral branches is little or nothing, as in woods and groves, then the top branches will mightily prevail; but when in a free open air, the perspiration and attraction of the lateral branches comes nearer to an equality with that of the top, then are the aspirings of the top branches greatly checked. And the case is the same in most other vegetables, which when they stand thick together, grow much in length with very weak lateral shoots.

And as the leaves are thus serviceable in promoting the growth of a tree, we may observe that nature has placed the pedals of the leaves-stalks where most nourishment is wanting, to produce leaves, shoots and fruit; and some such thin leafy expansion is so necessary for this purpose, that nature provides small thin expansions, which may be called primary leaves, that serve to protect and draw nourishment to the young shoot and leaf-buds before the leaf itself is expanded.

And herein we see the admirable contrivance of the Author of nature in adapting her different ways of conveying nourishment to the different circumstances of her productions. For in this embrio state of the buds a suitable provision is made to bring nourishment to them in a quantity sufficient for their then small demands: But when they are in some degree increased and formed, a much greater quantity of nourishment, is necessary, in proportion to their greater increase: Nature, that she may then no longer supply with a scanty hand, immediately changes her method, in order to convey nourishment with a more liberal hand to her productions; which supply daily increases by the greater expansion of the leaves, and consequently the more plentiful attraction and supply of sap, as the greater growth and demand for it increases.

We find a much more elaborate and beautiful apparatus, for the like purpose, in the curious expansions of blossoms and flowers, which seem to be appointed by Nature not only to protect, but also to draw and convey nourishment to the embrio fruit and seeds. But as soon as the *Calix* is formed into a small fruit, now impregnated with its minute seminal tree, furnished with its Secondine, *Corion* and *Amnion*, (which new set fruit may in that state be looked upon as a compleat egg of the tree, containing its young unhatched tree, yet in embrio) then the blossom falls off, leaving this new formed egg, or first set fruit in this infant state, to imbibe nourishment sufficient for it self, and the Fœtus with which it is impregnated: Which nourishment is brought within the reach and power of its suction by the adjoyning leaves.

If I may be allowed to indulge conjecture in a case, in which the most diligent inquirers are as yet, after all their laudable researches, advanced but little farther than meer conjecture, I would propose it to their consideration, whether from the manifest proof we have that sulphur strongly attracts air, a hint may not be taken, to consider whether this may not be the primary use of the *Farina fœcundans*, to attract and unite with it self elastick or other refined active particles. That this *Farina* abounds with sulphur, and that a very refined sort, is probable from the subtle oil which chymists obtain from the chives of saffron. And if this be the use of it, was it possible that it could be more aptly placed for the purpose than on very moveable *Apices* fixt on the slender points of the *Stamina*, whereby it might easily with the least breath of wind be dispersed in the air, thereby surrounding the plant, as it were, with an Atmosphere of sublimed sulphureous pounce? for many trees and plants abound with it, which uniting with the air particles, they may perhaps be inspired at several parts of the plant, and especially at the *Pistillum*, and be thence conveyed to the *Capsula seminalis*, especially towards evening, and in the night when the beautiful *Petala* of the flowers are closed up, and they, with all the other parts of the vegetable, are in strongly imbibing state. And if to these united sulphureous and aereal particles we suppose some particles of light to be joyned, for Sir *Isaac Newton* has found that sulphur attracts light strongly, then the result of these three by far the most active principles in nature, will be a *Punctum Saliens* to invigorate the *seminal* plant: And thus we are at last conducted, by the regular Analysis of vegetable nature, to the first enlivening principle of their minutest origin.

The Conclusion.

We have from the foregoing Experiments many proofs of the very great and different quantities of moisture imbibed and perspired by different kinds of Trees, and also of the influence the several states of the air, as to warm or cold, wet or dry, have on that perspiration. We see also what stores of moisture nature has provided in the Earth against a dry season, to answer this great expence of it in the production and support of vegetables; how far the dew can contribute to this supply, and how insufficient its small quantity is towards making good the great demands of perspiration: And that plants can plentifully imbibe moisture thro' their stems and leaves as well as perspire it.

We see with what degrees of warmth the sun, that kindly natural genius of vegetation, acts on the several parts of vegetables, from their tops down to their roots two feet under ground.

We have also many proofs of the great force with which plants and their several branches and leaves imbibe moisture, up their capillary sap vessels: The great influence the perspiring leaves have in this work, and the care nature has taken to place them in such order, and at such proper distances, as may render them most serviceable to this purpose, especially in bringing plenty of nourishment to the young growing shoots and fruit whose stem is usually surrounded with them near the fruit's insertion into the twig.

We see here too that the growth of shoots, leaves and fruit, consists in the extension of every part; for the effecting of which, nature has provided innumerable little vesicles, which being replete with dilating moisture, it does thereby powerfully extend, and draw out every ductile part.

We have here also many instances of the great force of the ascending sap in the vine in the bleeding season; as also of the sap's freely either ascending or descending, as it shall happen to be drawn by the perspiring leaves; and also of its ready lateral motion thro' the laterally communicating sap vessels; together with many proofs of the great plenty of air drawn in and mixed with the sap and incorporated into the substance of vegetables.

If therefore these Experiments and Observations give us any farther insight into the nature of plants, they will then doubtless be of some use in Agriculture and Gardening, either by serving to rectify some mistaken notions, or by helping farther to explain the reasons of many kinds of culture, which long repeated experience has found to be good, and perhaps by leading us to make some advances therein: But as it requires a long series and great variety of frequently repeated Experiments and Observations, to make a very small advance in the knowledge of the nature of vegetables; so proportionably we are from thence only to expect some gradual improvements in the culture of them.

The specifick differences of vegetables, which are all sustained and grow from the same nourishment, is doubtless owing to the very different formation of their minute vessels, whereby an almost infinite variety of combinations of the common principles of vegetables is made; whence some abound more with some principles and some with others. Hence some are of a warmer and more sulphureous, others of a more watry, saline, and therefore colder nature; some of a more firm and lasting, others of a more lax and perishable constitution. Hence also it is that some plants flourish best in one climate, and others in another; that much moisture is kindly to some, and hurtful to others; that some require a strong, rich, and others a poor, sandy soil; some do best in the shade, and others in the sun, *&c.* And could our eyes attain to a sight of the admirable texture of the parts on which the specifick differences in plants depend, what an amazing and beautiful scene of inimitable embroidery should we behold? what a variety of masterly strokes of machinery? what evident marks of consummate wisdom should we be entertained with?

We may observe that the constitution of plants is curiously adapted to the present state of things, so as to be most flourishing and vigorous in a middle state of the air, *viz.* when there is a due mixture and proportion of warm and cold, wet and dry; but when the seasons deviate far to any extream of these, then are they less or more injurious to the several sorts of vegetables according to the very different degrees of hardiness, or healthy latitude they enjoy.

The different seasons in which plants thrive best, seems to depend, among other causes, on the very different quantities imbibed and perspired by different kinds of plants. Thus the Ever-greens perspiring little, and having thereby a thick, viscid, oily sap, they can the better endure the winter's cold, and subsist with little fresh nourishment: They seem many of them to flourish most in the temperate seasons of the year, but not so well in the hottest part of Summer, because their perspiration is then somewhat too great, in proportion to the slow ascent of the sap, which makes some of them at that season to abate of their vigor: Thus some plants, which grow and thrive with the slow perspiration of *January and February*, perish as the spring advances, and the warmth and perspiration is too great for them. And thus garden Pease and Beans, which are sown in what is found to be their proper season, *viz.* in *November*, *January*, or *February*, tho' they make but a slow progress in their growth upwards, during the cold season, yet their roots, as also those of winter corns, do in the mean time shoot well into the warmer Earth, so as to be able to afford plenty of nourishment when the season advances, and there is a greater demand of it both for nutrition and perspiration. But when Pease are sow in *June*, in order for a crop in *September*, they rarely thrive well, unless in a cool moist summer, by reason of the too great perspiration caused by the summer's heat, which drys and hardens their fibres before they are full grown.

Tho' we have from these Experiments, and from common observation, many proofs of the great expansive force, with which the fibrous roots of plants shoot, yet the less resistance these tender shoots meet with, the greater progress they will certainly make in equal times: And therefore one considerable use of fallowing and trenching ground, and of mixing therewith several sorts of compost, as Chalk, Lime, Marle, Mold, *&c.* is not only thereby to replenish it with rich manure, but also to loosen and mellow the soil, not only that the air may the more easily penetrate to the roots, but also that the roots may the more readily make vigorous shoots. And the greater proportion the surface of the roots bears to the surface of the plants above ground, so much the greater quantity of nourishment they will afford, and consequently the plants will be the more vigorous, and better able to weather it out, against unkindly seasons, than those plants whose roots have made much shorter shoots. Herein therefore consists the great care and skill of the Husbandman, to adapt his different sorts of Husbandry to the very different soils, seasons and kinds of grain; that the several sorts of earth, from the very stiff and strong ground, to the loose light earths, may be wrought to the best temper they are capable of, for the kindly shooting and nourishing of the roots. And probably the Husbandman might get many useful hints, to direct him in adapting the several kinds of manure, and different sorts and seasons of culture to his different soils and grains: If in the several stages and growth of his Corn, he would not only make his observations, on what appears above ground, but would also frequently dig up, compare and examin the roots of plants of each sort, especially of those which grew in different soils, and were any how cultivated in a different manner from each other; this would inform them also, whether they sowed their Corn too thick or too thin, by comparing the branchings and extent of each root, with the space of ground allotted it to grow in.

And since we find so great a quantity of air inspired and mixt with the sap, and wrought into the substance of vegetables, the advantage of ploughing and fallowing ground seems to arise not only from the killing the weeds, and making it more mellow, for the shooting of the roots of Corn; but it is thereby also the better exposed to have the fertilizing, sulphureous, aereal and acid particles of the air mixt with it, which make land fruitful, as is evident from the fertility which the sword or surface of land acquires, by being long exposed to the air, without any culture or manure whatever.

We have seen many proofs of the great quantities of liquor imbibed and perspired by plants, and the very sensible influence which different states of the air had on their more or less free perspiration: A main intention therefore to be attended to in the culture of them, is to take due care, that they be sown or planted in proper seasons and soils, such as will afford them their due proportion of nourishment; which soils, as they are exhausted, must, as 'tis well known, from time to time, be replenished with fresh compost, such as is full of saline, sulphureous and aereal particles, with which common dung, lime, ashes, sword, or burn-bated turf abound: As also such manures as have nitrous and other salts in them; for tho' neither nitre nor common salt be found in vegetables; yet since they are observed to promote fertility, it is reasonable to conclude, that their texture is greatly altered in vegetation, by having their acid volatile salts separated from the attracting central air and earthy particles, and thereby making new combinations with the nutritive juice; and the probability of this is further confirmed from the great plenty of air and volatile salt, which is found in another combination of them, *viz.* in the Tartar of fermenting liquors: For it is the opinion of Chymists, that there is but one volatile salt in nature, out of which all other kinds of salts are formed by very different combinations, all which nutritive principles do by various combinations with the cultivated earth, compose that nutritive ductile matter, out of which the parts of vegetables are formed, and without which the watry vehicle alone cannot render a barren soil fruitful.

Nor is this the only care, the thriving and fertility of plants and trees depends much upon the happy influence and concurrence of a great variety of other circumstances. Thus many trees are unfruitful by being planted too deep, whereby their roots being in too moist a state, and too far from the proper influence of the Sun, whose power greatly decreases the deeper we go, as we see in Experiment 20. they imbibe too much crude moisture, which tho' productive of wood, is yet unkindly for fruit.

Or if when not planted too deep, they are full of crude sap, either by being too luxurious, or too much shaded; or are planted in a moist, when they delight in a dry soil, then the sap is not so sufficiently digested by the Sun's warmth, as to be in that ductile state, which is proper for the producing of fruit.

And thus the Vine, which is known to thrive well in a dry, gravelly, rocky soil, will not be so fruitful in a moist, stiff, clay ground: And accordingly we may observe in Experiment the 3d, that tho' the Vine imbibed and perspired more than the Ever-green, yet it perspired less than the Apple-tree, which delights in, and bears best in a strong brick-earth clay; for tho' the Vine bleeds most freely in its season, produces many long succulent shoots, and bears great plenty of a very juicy fruit, yet from that Experiment it is plain, that it is not a great perspirer, and therefore thrives best in a dry, rocky, or gravelly soil.

The considerable quantity of moisture, which by Experiment 16. is perspired from the branches of trees, during the cold winter season, plainly shews the reason, why in a long series of cold north-easterly winds, the blossoms, and tender young set fruit and leaves, are in the early spring so frequently blasted, *viz.* by having the moisture exhaled faster than it can be supplied from the trees; for doubtless that moisture rises the slower from the root, the colder the season is, tho' it rises in some degree all the winter, as is evident from the same Experiment.

And from the same cause it is, that the leafy spires of Corn are by these cold drying winds often faded and turned yellow; which makes the Husbandman, on these occasions, wish for snow; which tho' it be very cold, yet it not only defends the root from being frozen, but also screens the

Corn from these drying winds, and keeps it in a moist, florid, supple state.

It seems therefore to be a very reasonable direction which is given by some of the Authors who write on Agriculture and Gardening, *viz.* during these cold drying winds, when little dew falls, to water the trees in dry soils, in the blossoming season, and while the young set fruit is tender; and provided there is no immediate danger of a frost, or in case of continued frost, to take care to cover the trees well, and at the same time to sprinkle them with water, which is imitating nature's method of watering every part: But if the success of this practice in cold weather may be thought a little doubtful; yet the sprinkling the bodies and leaves of trees, in a very hot and dry summer season, seems most reasonable, for by Exper. 42. they will imbibe much moisture.

As to sloping shelters over Wall-trees, I have often found, that when they are so broad as to prevent any rain or dew coming at the trees, they do more harm than good, in these long easterly drying winds; because they prevent the rain and dews falling on them, which would not only refresh and supple them, but also convey nourishment to them: But in the case of sharp frosts after showers of rain, these shelters and other fences must needs be of excellent use to prevent the almost total destruction which is occasioned by the freezing of the tender parts of vegetables, when they are full saturate with rain.

The full proof we have from these Experiments, of the serviceableness of the leaves in drawing up the sap, and the care we see nature takes, in furnishing the twigs with plenty of them, principally near the fruit, may instruct us on the one hand, not to be too lavish in pruning them off, and to be ever mindful to leave some on the branch beyond the fruit; and on the other hand, to be as careful to cut off all superfluous shoots, which we are assured do draw off in waste great quantity of nourishment. And might it not be adviseable, among many other ways which are prescribed, to try whether the too great luxuriancy of a tree or branch could not be much checked by pulling off some of its leaves? How many experience will best teach us, the pulling all off will endanger the killing the branch or tree.

There is another very considerable use of the leaves, *viz.* to keep the growing fruit in a supple ductile state, by defending it from the Sun and drying winds, which by toughning and hardening its fibres spoil its growth, when too much exposed to them; but when full grown, or near it, a little more Sun is often very needful to ripen it. In hotter climates fruits want more shade than in this country, and here too, more shade is needful in a hot dry summer, than in a wet cool one.

The consideration of the strong imbibing power of the branches of trees, and the readiness with which we see the sap passes to and fro, to follow the strongest attraction, may perhaps give some useful hints to the Gardiner, in the pruning and shaping of his trees, in checking the too luxuriant, and helping and encouraging the unthriving parts of trees.

It is a constant rule among Gardiners, founded on long experience, to prune weak trees early in the winter, because they find that late pruning checks them; and for the same reason to prune luxuriant trees late in the spring, in order to check their luxuriancy. Now it is evident that this check does not proceed from any considerable loss of sap at the wounds of the pruned tree, excepting the case of a few bleeding trees when cut in that season, but must arise from some other cause; for by Experiment 12 and 37. where mercurial gages were fixed to the stems of fresh cut trees, those wounds were constantly in a strongly imbibing state, except the Vine in the bleeding season.

When a weak tree is pruned early in the beginning of the winter, the orifices of the sap-vessels are closed up long before the spring, as is evident from many Experiments in the 1st, 2d and 3d chapters; and consequently when in the spring and summer the warm weather advances, the attracting force of the perspiring leaves is not then weakened by many inlets from fresh wounds, but is wholly exerted in drawing sap from the root. Whereas on the other hand, when a luxuriant tree is pruned late in the spring, the force of its leaves to attract sap from the root will be much spent and lost at the several fresh cut inlets.

Besides, the early pruned tree being eased of several of its twigs or branches, has thereby the advantage of standing thro' the whole winter, with a head better proportioned to its weak root. And since by Exper. 16. the sap is found to ascend in the winter, less of that than cold crude juice is drawn thro' the roots and stem, to supply the perspiration of the remaining boughs, whereby the sap of the tree is probably less depauperated than it would have been, if all the boughs had remained on. For these reasons, early pruning should in the main, and excepting some cases, be better than late.

And the reasonableness of this practice is further confirmed by the experience of Mr. *Palmer*, a curious Gentleman of *Chelsea*, who has found, that by pruning his Vines, and pulling all the leaves off them in *September*, as soon as the fruit was off, they have borne greater plenty of Grapes than other Vines, particularly in the year 1726. when by reason of the extreme wetness and coldness of the preceding summer, the unripe shoots produced generally very little fruit.

From many Experiments in the second Chapter, the Gardiner will see with what force his grafts imbibe sap from the stock, especially that ductile nourishment from between the bark and wood; which corresponding parts he well knows by constant experience must be carefully adapted to each other in grafting, those grafts being always best whose buds are not far asunder, *viz.* because their expanding leaves can therefore draw up sap the more vigorously.

The great quantities of moisture which we find by Experiment 12 are imbibed at wounds where branches are cut off, shews the reasonableness of the caution used by many who are desirous to preserve their trees, *viz.* either by plaistring or covering with Sheet-lead the very large wounds of trees, to defend their trunks from being rotted by the soaking in of rain.

And from the same 12th Experiment a hint may be taken to make some attempts to give an artificial taste to fruits, by making trees imbibe in the same manner some strongly tinged or perfumed liquor, which is not spirituous, for that we see will kill the tree. I have made the stem of a branch of a tree imbibe two quarts of water without killing it; If any are desirous to make this Experiment, they should take care to cut the stump which is to imbibe the liquor as long as they can, that there may be the more room, from time to time, to cut off an inch or two of the top, when it is grown to saturate with liquor that more will not pass.

Tho' Ever-greens are found to imbibe and perspire much less than other trees, yet is the quantity they perspire so considerable, that it has always been one of the greatest difficulties in the ordering of a Green-house to let in fresh air enough without exposing the plants to too much cold. For since the perspiration of trees will not be free and kindly in a close damp air, the sap will be apt to stagnate, which will make the plants grow moldy, or they will be sickly, by imbibing such damp rancid vapours; for by Mr. *Miller*'s curious observations on the perspiration of the *Plantain* tree of the *West-Indies*, and of the *Aloe* under *Experiment* 5, plants will often imbibe moisture in the night as well in stoves as common Green-houses without fire; it is certainly of as great importance to the life of the plants to discharge that infected rancid air, by the admittance of fresh, as it is to defend them from the extream cold of the outward air, which will destroy them if let in immediately upon them. It seems therefore to be a very reasonable method which some use, *viz.* to cover some of the inlets of their Green-houses on all sides with canvass, and in extream cold weather with shutters made of reed or straw, through which the air can only pass in little streams: The like contrivance would probably also be of good service to purify gradually the thick rancid fumes which arise from the dung of hot beds, and are often very destructive of the tender plants: This is to imitate nature, which while she provides for the defence of living creatures against the cold, by a good covering of Hair, Wool, or Feathers, at the same time she takes care that the air may have admittance through innumerable narrow meanders in such quantities as may be sufficient to carry off the perspiring matter.

I have here, and as occasion offered under several of the foregoing Experiments, only touched upon a few of the most obvious instances, wherein these kind of researches may possibly be of service in giving us useful hints in the culture of plants: Tho' I am very sensible, that it is from long experience chiefly that we are to expect the most certain rules of practice, yet it is withal to be remembred, that the likeliest method to enable us to make the most judicious observations, and to put us upon the most probable means of improving any art, is to get the best insight we can into the nature and properties of those things which we are desirous to cultivate and improve.

FINIS.

A TABLE where to find each EXPERIMENT.

A TABLE where to find each EXPERIMENT.

A TABLE where to find each FIGURE.